THE
CHAOS
SCENARIO

D1446078

by
Bob Garfield

The Chaos Scenario
Copyright © 2009 by Bob Garfield

Some of the material in this book, though revised and expanded, originated
in *Advertising Age*, AdAge.com and *Wired*. The *Ad Age* material is used by
permission of the copyright holder, Crain Communications.

Printed in the United States of America

09 10 11 12 13 14 15 • 15 14 13 12 11 10 9 8 7 6 5 4 3 2 1

For Milena

"If you don't like change, you're going to like irrelevancy even less."

— *General Eric Shinseki, 2003*

"Perhaps if the existing community would take now and then the trouble to pass in review the changes it has already witnessed, it would be less astounded at the revolutions which continually do and continually must flash before it; perhaps also it might with more grace accept the inevitable, and cease from the useless attempts at making a wholly new world conform itself to the rules and theories of a bygone civilization."

— *Charles Francis Adams Jr., contemplating the transcontinental railroad, 1868*

"Garfield, will you ever finish that book?"

— *Milena Garfield, 2005, 2006, 2007, 2008, 2009*

CONTENTS

THE ART AND SCIENCE OF LISTENOMICS

THE DEATH OF EVERYTHING

The digital world has so disrupted the business models of newspapers, radio, television, music and even Hollywood that the yin and yang of mass media and mass marketing are flying apart. We are in the midst of total collapse of the media infrastructure we have taken for granted for 400 years.

THE POST-ADVERTISING AGE

A digital universe, in the fullness of time, will emerge as an ideal environment for marketing. But that won't likely be supported by advertising; ad avoidance, ultra-fragmentation and the pesky Law of Supply and Demand conspire the eradicate advertising's primacy. In the short run, this bodes very poorly for new episodes of *Lost*.

THE WORLD'S MOST SUCCESSFUL FAILURE?

YouTube is a global phenomenon that has altered human behavior on a grand scale. It is also a big money loser, with dubious hope of ever being profitable. So now what?

THE ART AND SCIENCE OF LISTENOMICS

TODAY, MAY 3, 2009, as I sat in a Philadelphia chemistry lab conducting my final interview for this book, The New York Times Co. announced it would file a notice commencing the shutdown of its second largest news property, the *Boston Globe*.

After 137 years.

We shall see whether this was just a corporate gambit to win more union concessions, but anyway the typewriting is on the page. Sooner or later (most likely sooner), the thrumming presses of another great newspaper — this one arguably the democratic lifeblood of the whole of New England — will grind to a halt. So where to place the blame?

No mystery there. They were the same brutes responsible earlier in the year for bankrupting the *Chicago Tribune* and the *Minneapolis Star Tribune* and the *Philadelphia Inquirer*. For shuttering the *Rocky Mountain News*, and *Portfolio* magazine and *Playgirl* and *Domino*. For ravaging the music industry. For decimating the broadcast industry. And for throttling the advertising industry to within a column inch of its life.

The culprits: zeros and ones.

See, the thing is, "The Digital Revolution" isn't just some newsmagazine cover headline. It's an *actual* revolution, yielding revolutionary changes, thousands or millions of victims and an entirely new way of life. It's not just that you can talk to your refrigerator or bank online or E-ZPass your way through the toll booth while those other suckers in the right lanes are backed up clear to that horrible rest stop with the price-gouging Sunoco and ammonia-scented Sbarro. Those are just minor conveniences afforded by the very same binary code fueling the real conflagration. Maybe you've been too busy fiddling with your smart phone to notice, but the mass media and mass marketing structures that have more or less defined your connection with the world for more than a century are in flames.

As you shall see in the first chapter, "The Death of Everything," the times they are a changin'. Traditional media are in a stage of dire retrenchment as prelude to complete collapse. Newspapers, magazines and especially TV as we currently know them are fundamentally doomed, as they shudder against three concurrent, irresistible forces: 1) audience shrinkage with consequent advertiser defection, 2) obsolete methods — and unsustainable costs — of distribution and 3) competition from every computer user in the whole wide world. What you call articles and TV shows and songs, and what the media industry calls "content," will never be the same again. This will change your media environment in dramatic ways. It will change the advertising industry in melodramatic ways.

Madison Avenue, after all, exists to create ads to subsidize the vast expense along the vast expanse of old media. It has footed the bill for *Gilligan's Island*, *The New Republic*, *The Family Circus*, *Rush Limbaugh*, *TRL*, and *The Wall Street Journal* not for the fun of it, but because marketers depended on those media to reach mass audiences. Indeed, they've paid increasing premiums for the opportunity as audiences have shrunk, because even in a fragmented media world, the largest fragment — network TV — is the most valuable. Now they realize that they are losing not only mass but critical mass. When that is gone, marketers have no reason to advertise, no remotely similar place to spend that $47 billion a year. Therefore, as culturally improbable as it may sound, the days of Madison Avenue dictating messages to you are all but at an end.

Goodbye, Mr. Whipple. Fondle all the Charmin you want, but do it on your own time.

Mind you, I'm not talking about the death of marketing and media. I'm talking about a dramatic rebirth in marketing and media, in approximately the way the end of the last Ice Age yielded exponentially more species, and more advanced species, than had ever prospered on earth. When the TV Age finally succumbs to the Digital Age, we will be living a different world. And (mainly) a much better one. But for those entrenched in the status quo, involuntary change can be a difficult concept to accept. Titans of the Old Model have spent the past five years wallowing in various forms of hubris and denial. You'll see some examples in Chapter 2 ("The Post-Advertising Age") but for the moment let me just share the 2007 comments of Sir Martin Sorrell, chairman of the WPP Group, the world's largest advertising agency holding company:

"Slowly, the new media will cease to be thought of as new media; they will simply be additional channels of communication. And like all media that were once new media but are now just media, they'll earn a well-deserved place in the media repertoire, perhaps through reverse takeovers — *but will almost certainly displace none.*"

The italics are mine. The absurdity was Sir Martin's. Does he not see that the internet is not just some newfangled medium — like TV displacing radio? No, it is a revolutionary advance, along the lines of fire, agriculture, the wheel, the printing press, gunpowder, electricity, radio, manned flight, antibiotics, atomic energy and, maybe, Listerine breath strips. The digital revolution is already having far-ranging effects on every aspect of our lives, from socialization to communication to information to entertainment to democracy, and these Brave New World effects will only be magnified as the Cowardly Old World collapses before our eyes. Not that this *will* happen.

This *is* happening. Right now.

So, here's a thought: why not spend the better part of four years documenting the destruction of the old mass-media/mass-marketing business model, while simultaneously envisioning the micro-media/ micro-marketing world that will succeed it? Splendid idea. Two! Two! Two books in one!

The Chaos Scenario is about the historic re-ordering of media, marketing and commerce triggered by the revolution in digital technology. It will explore five continents for examples of adaptation to what is literally a new age of human endeavor. It is about the cutting edge, which is sometimes a laser scalpel but, of course, also sometimes a guillotine. It is, in short, about crawling from the wreckage of the old order to establish a new one. So let us therefore begin in that notable crucible of apocalyptic disruption — Billund, Denmark — at a company that makes plastic toy bricks.

Billund (pop. 8697) is a squat patch of agricultural land square in the center of the Jutland Peninsula, which does its jutting forlornly northward from Germany. In the springtime, after the thaw, stepping off a turbo-prop produces an instant reminder of how rural the place is. The earthy perfume of manure shoots right up your nose, as if you'd thumbed past a *Vogue* ad sampling Merde from Estée Lauder. Let's just say Billund isn't Bangalore or the Silicon Valley. It does not immediately announce: "technology."

From the airport it is but a mile to Hotel Legoland., which itself will remind nobody of, say, the Taj Palace Dubai. It's more like a sprawling, Lego-themed Best Western, at the driveway to which visitors are greeted by a gigantic animatronic bellman made of Lego bricks, grinning and saluting, grinning and saluting, like a fascist Big Boy. In the hotel, corridors with such fanciful names as Fairytale Road and City Avenue are lined with other colorful Lego constructions. The hallway to my room was guarded by a 4-foot-tall gnome wielding a sign that announced, a bit hyperbolically, "Castle Street." If you were little Dorothy, you would turn to your dog and say, "Toto, I don't think we're in Kansas anymore."

But you would be wrong. Billund is very much like Kansas, only with herring for breakfast. The natives are hardy and stolid, kind but reserved — and maybe even a bit flinty. Rural Denmark is not a place given much to effusion or showing off. It is a place where folks get out of bed in early morning and go about their business, then return in mid-afternoon to their families, whom presumably they feel no need to impress with worldliness or thought leadership. In fact, if you were to just stroll around this eccentric Polymer Motor Inn taking in the various remarkable examples of plastic-brick statuary, you might even be struck by a further hint of backwardness. In a world of "World of Warcraft" and Second Life, the Lego figures — no matter how elaborate — are nonetheless quaintly retro. Whether assembled as bellboys, gnomes, Darth Vader, cheerful workmen or — as in the lobby art — the Mona Lisa, their constituent rectangular bricks are like 3-D plastic pixels. Everything is just so.... low-resolution.

So, yes, Billund is the birthplace each year of millions upon millions of old fashioned, low-tech building blocks molded in a cow pasture. It is also where I began my journey as a chronicler of revolution. For here the future of commerce is being forged.

Various Kinds of Omics

First for its line of robotic toys called Mindstorms, and now expanding to its Lego Creator and Lego Factory lines, this company has pioneered the notion of consumer involvement — not only soliciting ideas from its most loyal and enthusiastic customers, but actively recruiting them for product design. Mindstorms, which first appeared in 1998, was itself a forward-looking enterprise for a company that began in 1932 selling wooden pull toys, transitioned to colorful plastic bricks in 1949 and basically stood pat for 50 years. But the original Mindstorms technology was complicated and sales slow. That is, until 2002, when the Mindstorms User Group stepped off their turbo-prop, sucked in some methane, and got to work on the product's second generation. For 14 months, between Billund and their own home computers, these volunteers reinvented the brand — which now is a soaring success. And not only did they fly to Jutland at their own expense for the privilege of being unpaid consultants, they turned right around and evangelized the resulting products to the greater community of fans, geeks and Lego freaks. Internet fan sites, in no way contributed to or controlled by the company, represent virtually the entire Mindstorms marketing program.

The Mindstorms experiment took place in the midst of a crisis at Lego. After the millennium, as electronic toys and internet games increasingly usurped their customers' attention, sales flattened and profitability disappeared. In 2004, losses were so steep, the company was in danger of liquidation. That crisis, perhaps combined with revelations about the passion and commitment of the core audience emerging from the Mindstorms project, led management to rethink every aspect of its business, from dumping extraneous product lines to trimming the workforce to institutionalizing the consumer-participation concept. Under the new management structure, "community education & direct" is one of only four lines of authority within the company — co-equal with administration, supply-chain management and sales & marketing. Its function: to deal directly with consumers, whose collective wisdom, enthusiasm and judgment — as demonstrated in forum after forum online — exceeds that of the company itself.

What this enlightened organization is doing, in other words, is listening. To loyal consumers, to dissatisfied consumers, to employees, to suppliers, to any faint echo in the marketplace that may help it sell plastic bricks to the world. This exercise doesn't necessarily have to take place by flying people to headquarters. It can go on every second of the day, on an organization's own websites and on the websites of others. And it will create connections, data and insights such as you never enjoyed before, or perhaps even imagined before. Because it turns out that all those guys with the PowerPoint presentations you've been sitting through for the past three years — you know, the ones insisting that "the consumer is in control" — are absolutely right. The consumer (and voter and citizen) is in control: of what and when she watches, of what and when she reads, of whether to pay any attention to you whatsoever or to make your life a living hell. This might be an excellent time, therefore, to listen to what she has to say. And it sure wouldn't hurt to make her your friend.

This is the future of everything. In fact, if you wish to survive for long in media, marketing, politics or any other institution accustomed to managing its affairs from the top down, it is the *right now* of everything. Survival means institutionalizing dialogue with all of your potential constituencies and sometimes total strangers for the purpose of market research, product development, customer relationships, corporate image and transactions themselves. The last of those benefits is especially important, because when you sell goods or services, you get money.

That is the essence of "Listenomics," my term for the art and science of cultivating relationships with individuals in a connected, increasingly open-source environment. It is also what this book probably should be titled. Unfortunately, I was pre-empted by *Wikinomics* and freakin' *Freakonomics*, two fine books that went all -omic on the publishing world before I got a chance to. So while it may not be such a unique coinage, it is a fitting discipline for the most extraordinary world in which it will flourish — a world very unlike anything we've experienced before.

"The era of the creepy blue light leaking out of every living room window on the block is now officially at an end," says my pal and occasional colleague Steve Rosenbaum, founder of video-sharing startup Magnify.net and arguably the inventor a decade ago of citizen video. "The simple, wonderful, delirious fact is that people like you and me can now make and share content."

Steve refers to blogs. He refers to consumer-generated commercials, such as the Doritos ads that appeared on the Super Bowl. He refers to pop songs self-produced and distributed by garage bands, who suddenly have as much access to an audience as Madonna. But mostly he refers to video.

Think about YouTube. In 2005, it did not exist. Now it hosts millions of video clips of every description, many created out of whole pixels by ordinary civilians. It feeds 200 million clips a day, most to MySpace, Facebook and other social networks, and has supplanted MTV as the prime media destination for the Short-Attention-Span Generation. Google bought it for $1.65 billion and Viacom is suing it for $1 billion — and both for the same reason: it has utterly disrupted the status quo of audience behavior, content distribution and Hollywood's hitherto impregnable monopoly on making stuff people get to watch. But here's a fun fact. YouTube (Chapter 3. "The World's Most Successful Failure?") doesn't sell enough advertising to cover even 20% of its bandwidth bill, because 1) online ads don't fetch much money, and 2) the audience doesn't want any part of them.

Bear in mind, as decades of data prove, consumers have never much cared for advertising. They accepted it only because, apart from being their part of the deal for cheap or free content, it was basically unavoidable. Now, however, an entire generation has grown up getting free content online without much advertising interruption, and they consider it their birthright to do so. Moreover, they now have the technology — such as TiVo, spam blockers, etc. — for avoiding advertising, which is exactly what they do. For that reason, and because of the collapse of the mass-media yin bodes poorly for mass-marketing's yang, advertising does not have a very bright future. This most likely spells a protracted nightmare for the ad-agency business, which at the moment expresses equal parts panic and Sorrellian denial. But if you happen to be in the business of flogging goods — or policies or laws — fear not. All is not lost. Salvation is within your grasp. Just listen.

The Great Mississippi Reverses Flow

Can you hear it? In the distance? It's a crowd forming — a crowd of what you used to call the "audience." They're still an audience, but they aren't necessarily listening to you. They're listening to each other talk about you. And they're using your products, your brand names, your iconography, your slogans, your trademarks, your designs, your goodwill, all of it as if it belonged to them — which, in a way, it all does, because, after all, haven't you spent decades, and trillions, to convince them of just that? Congratulations. It worked. The Great Consumer Society believes deeply that it has a proprietary stake in you. Remember New Coke fiasco? Epic "fail." Having spent millions of dollars developing a new formula, and millions more doing consumer research that proved beyond dispute that consumers preferred the sweeter formulation compared to the classic, century-old Merchandise 7X, in 1985 the Coca Cola Co. sprang New Coke on the world only to discover — to its astonishment and horror — that the world wasn't amused. Yes, the company had done blind taste tests to a fare-thee-well, but "Which do you prefer?" was not the right question. The right question was: "Do you want us to change the Coke formula?" Had they bothered to inquire, they would have discovered the answer to be an emphatic no. The consumers not only were emotionally attached to the old flavor, they regarded themselves as having a proprietary interest in the product. They had drunk so much Coke, they believed it in a very real sense belonged to them. And like stakeholders everywhere, they let their voices be heard.

That was a quarter century ago. Now imagine that phenomenon, magnified exponentially, via the internet.

"I think it's going to be more and more of an open conversation, as opposed to dictation," says internet guru Battelle, a founder of *Wired* and *The Industry Standard.* "Marketers are increasingly going to have to adopt the principles of the environment in which they find themselves."

Why? Because, as I've been explaining, the information society is reversing flow. We are increasingly inhabiting an open-source world. What began as an experiment among a few software nerds has, thanks to the internet, expanded into other disciplines, notably media and law. But it won't stop there. Advertising. Branding. Distribution. Consumer research. Product development. Manufacturing. They will all be turned upside down as the despotism of the executive suite gives way to the will and wisdom of the masses in a new commercial and cultural epoch. As the people at Lego have figured out, and Sir Martin Sorrell evidently has not, is that the post-advertising age is The Listenomics Age. Its defining characteristic: the herd will be heard. If you do not listen carefully, you are a fool. Not because the crowd is a threat (although, of course, it is) but because it is your greatest resource. What if its wisdom were harnessed and its power unleashed, unfettered by outmoded intellectual-property laws and uninhibited by the dictates of Management? Here's what: payday.

And this, too, is already in progress. For example, as augured by the New Coke fiasco, the number two Facebook fan page — after Barack Obama's — is for Coca-Cola. It wasn't created by the company; it was created by two fans and joined by 3.3 million more while the brand stewards in Atlanta just watched from a distance with their jaws agape. In this book you will be introduced to many ways in which the simple exercise of listening enhances and even replaces business disciplines that have undergirded commerce since time immemorial. In Chapter 4 ("Talk is Cheap") you'll learn how the oldest communication channel in the world — word of mouth — has been supercharged by the internet, rendering much traditional marketing irrelevant on credibility grounds. You'll travel to Tel Aviv, where a young mathematician/psychologist writes algorithms to plumb the meanings, and influence, of words he will never lay eyes on. Chapter 6, "Comcast Must Die," discusses how mobilized consumers have used their newfound leverage to embarrass, shame and potentially even cripple vast corporations — including the woeful cable giant who made the fatal mistake of pissing me off. In that chapter, I'll also discuss the proliferation of product reviews, on such sites as Angie's List and ePinions.com, and hate sites where all the hatred is targeted at a commercial brand. In Chapter 7, ("Guess") I'll discuss how monitoring the data of online transactions — a la Netflix and Amazon.com — can become the core of a business. In Chapter 9 ("Off, Off, Off Madison"), you'll learn about consumer-generated ads like those you've seen on the Super Bowl. Then, of course, there is Chapter 8 ("Sometimes You Just Gotta Lego"); about Danish toymakers whose colorful plastic bricks, like so many Homeric sirens, lure fanboys and their unbridled passions. You'll encounter equally revolutionary examples from journalism to encyclopedias to religion to electoral politics to political revolution, each demonstrating a hitherto unimaginable means of getting business done. This book is about all the many ways Listenomics can and must be incorporated into every organization — from Unilever to the United States government — that depends on the public for its sustenance.

Not that it will all be a bed of roses, of course. When you listen carefully, sometimes you get an earful. For every enthusiast who comes up with a new design for a Lego ferry boat there are millions of unenthusiasts who wish to weigh in:

VWsucksass.com, to cite one vivid example. .

Or, to pick on the same unfortunate marketer, one of the first consumer-generated ads was: a spoof commercial for the subcompact Polo. It featured a Palestinian suicide bomber trying to blow up a café. But the bomb causes no damage because Polo is "Small but tough." The fake ad got 12 million hits, but VW did not send a thank-you note. It threatened a lawsuit. Nothing much came of that. You can rattle lawyers all you want, but you can't take away everybody's computer. Once again, you are not in control of your message, your image or your reputation. The Consumer is — and there are a lot of her.

Note the capital C. Because, after all, the Consumer is everyone — which gets to a rather chilling aspect of Listenomics: the disturbing fact that it can only exist in a Brave New World. As Aldous Huxley predicted, the digital universe is essentially a total surveillance society. This can and will produce a number of chilling and undesirable effects, from Chinese info-fascism to cyber vigilantism and online reputation destruction. (Chapter 11. "Nobody is Safe from Everybody.") That phenomenon is a classic demonstration of the digital age's double-edged sword. Some of those caught in web of the web are innocent. Some are perhaps guilty of private indiscretions, but suddenly exposed to public opprobrium. Some are bona fide villains, like the sexual predators snared by the group Perverted Justice. And some are politicians, who will learn the hard way that the bully pulpit can suddenly expose them to the ire of the bullied.

For the ultimate lesson on that subject, we need look no farther than the plight of George Allen (Chapter 10. "The Powers That Be 2.0"). Not long ago, he was among the frontrunners for the Republican nomination to be president of the United States. Now he is the former senator from Virginia. He achieved this dubious distinction in the midst of his own putative exercise in Listenomics. This was his "Listening Tour" of Virginia, the centerpiece of his 2006 re-election race against Democratic longshot James Webb. At one stop on his tour, in rural Breaks, Virginia, candidate Allen paused from his listening to start running his mouth. In particular, he was blathering about a young man videotaping his every word with a digital camera — a young man Allen well knew to be in the employ of the Jim Webb campaign. He had been shadowing Allen from stop to stop along the tour, on the off chance that Allen would do or say something controversial and give the struggling challenger a straw to grasp in the race.

Of course, nobody being filmed for that purpose would be stupid enough to provide ammunition for the opponent — nobody, that is, except George Allen. Instead of ignoring the video pest, Allen made direct reference to him. "This fellow here over here with the yellow shirt: 'Macaca,' or whatever his name is...." Then he called him Macaca a second time. Then a third.

Nobody knows exactly why Allen should have chosen to call the guy such a name. What we do know is that the young fellow was a very dark-complected Indian-American, and that in North Africa, where Allen's mom is from, *macaca* is an epithet against black people. Oh, yeah, we know one other thing: Allen blew a 16-point lead and lost the election.

Because, see, here's the thing about Listenomics: Other people are listening back.

Chapter 1

THE DEATH OF EVERYTHING

IT'S A SUMMER EVENING IN MONTENEGRO. The skiffs in the marina sway gently as the sun prepares to set on the Adriatic, the salt water lapping against the ancient stone. The encircling mountains are mere shadows in the gloaming. On the quay, a lonesome cello plays a melancholy love song. I've come here for the sea, but I've blundered into a metaphor.

And several dizzying paradoxes. This is Budva, a medieval port in a brand-new country, thick with Serbian, Italian and Russian tourists for the August high season. Here worlds and ages seem ever to overlap. Narrow cobbled streets clogged with BMWs. A brazenly corrupt tax haven rising from the ashes of Communist Yugoslavia. Weary poverty mingling with splashy new money. Across the crescent harbor, hard by the *Citadela* — the fortress that guarded this Balkan paradise for centuries — bolted to an 800-year-old wall is a 600-square-foot LED tele-screen. It is showing advertisements, one after, another, 24 hours a day. But, as somebody once said, wait — there's more. On the far side of that wall, atop a shale cliff that plunges 50 feet to the sea, *sits Izmedu Crkava* — "between churches" — a public square nestled among the worshippers. On this spot, in a few hours, will commence *Noc Reklamozdera*.

The Night of Advertising. Tourists will buy tickets to see 90 minutes of TV commercials. It's the brainchild of a Frenchman named Jean Marie Boursicot, who's been staging such events all over Europe for years: a reel of ads followed by a dance party with a DJ. So there's another strange contrast for you. We can assume that some percentage of the audience has a DVR at home, with which they fast forward right through the very commercials they are now spending cash money to see.

But never mind even that little irony. Listen. Here at the harbor, I'm struck by that haunting cello, played by a pale, pretty busker to her indifferent passersby. What is that mournful melody? Ah, it's "Katyusha," the Russian "Lily Marlene." It's about a young girl crying for her lost love, while all around her everything blossoms. It's also what Russian armor soldiers sang as they they stuffed their tank barrels with shells and fired destruction through the sky. More fitting notes have never been bowed. Here in Budva, amid the perpetual clash of civilizations, they have only just buried the Communist revolution, yet they will celebrate "The Night of Ads" utterly oblivious to the fact that they are in the midst of a capitalist one. A love affair is over, and shells fly through the air on their way to destroy even the most fortified structures of the analog universe. This is a going-away party for commerce's old world order. Never mind the DJ's pounding techno. Listen to the cello. It is right there. Listen, and face the music.

Mass Exodus

Why, all of a sudden, is it so important to listen? Here's why: Because hardly anyone anymore is listening to you.

There was a time, essentially the six centuries since Gutenberg gave the world moveable type, when various political, clerical and commercial elites could speak to the masses and feel confident of having an attentive audience. For the past four centuries, mass media were funded or at least subsidized by mass marketing, which piggybacked on what we now call "content" to issue messages of its own.

Like the eternal co-dependence of flowers and bees, this was an extremely convenient symbiotic relationship for those involved. Or if you prefer a more spiritual analogy, imagine the media yin coupled snugly with the advertising yang, a transcendent oneness yielding cheap or free content for all. Well, that's over — or damn near. In the digital age, that time-honored symbiosis is coming apart. It's happening slowly enough that most consumers haven't really noticed. But it's happening quickly enough that media and marketing are in big trouble — trouble that I believe will send the world spinning into a postapocolyptic post-advertising age. In this chapter and the one following, I intend to prove that to you. Meantime, just think about what's happening all around us.

Have you given any thought, for instance, to the attack of the pod people?

Surely you've noticed them, on the subway or at the gym, all those folks milling about with little white buds in their ears. This is not a hygiene problem. This is *prima fascie* evidence of the ongoing revolution. These people may seem placid enough, but as they pump away at the elliptical machine or stare into the middle distance avoiding eye contact with their fellow straphangers, they are actually storming the Bastille. As they privately groove to digital recordings of U2 or Lil Wayne or *Ella Sings Cole*, they are simultaneously dismantling the Old World Order. Thanks to the iPod, the record business and commercial broadcast radio are *in extremis.*

Let's think about radio for a moment. Not only has satellite radio siphoned off audience from terrestrial stations, iTunes, et al, have rendered the random, maddeningly limited play lists generated by Z104 very pale competition. Those pod people are storming the Bastille essentially by virtue of bypassing three entire industries: radio, recording and advertising — the latter of which till now has underwritten almost everybody's music-listening desires. See, you don't have to listen to ads on your iPod. Even public radio is at risk, because it no longer holds a monopoly on programming and distribution of in-depth audio news and information. Sure, NPR now sends out much of its content via podcast, but so can anybody else. What NPR has that Joe Citizen doesn't have is hundreds of millions of dollars of increasingly unneces-sary costs — because nobody needs transmitters and gigantic broadcast towers to receive a stream or podcast. More nettlesome still, as broad-band penetration approaches universality, nobody will need a radio to hear what NPR feeds. Yet local stations, in their pledge drives, generate 95% of the network's money. Hmm.

Newspapers are in an even more precipitous downward spiral. They, too, are losing audience rapidly, and will never replace those readers, because just as young people no longer listen to the radio, they sim-ply don't buy newspapers. Circulation declines mean loss of display-advertising revenue, which, of course, is linked directly to audience. Meantime, classified advertising — hitherto the most profitable seg-ment of the entire media industry — has been rendered irrelevant by Craigslist, Monster.com, AutoTrader.com and eBay. The *Minneapolis Star Tribune*, acquired by McClatchy in 1998 for $1.2 billion, was sold to private investors in December 2006 for $530 million. In January 2009 it declared Chapter 11 bankruptcy. In the year 2000, the Chicago-based Tribune Co. was valued at $12 billion. It then bought the Times-Mirror Co. for more than $8 billion. Then, in April 2007, real estate devel-oper Sam Zell charged in as a white knight, and commenced wholesale retrenchment, including layoffs, bureau closings and the sale of *News-day*. Worked like a charm. Within 20 months, the Tribune company was in bankruptcy.

At this writing, A.H. Belo, owner of the *Dallas Morning News*, has laid off 25% of its workforce in the past year. The *Rocky Mountain News* folded in March 2009, and the *San Francisco Chronicle* and *Seattle Times* were on death's door. In 2008, Rupert Murdoch plunked down $5.5 billion last year for a $3.5 billion paper, the *Wall Street Journal*, and nobody knew whether he was a genius of synergy and valuation or a sucker. The recession obscures the answer, but in February 2009 News Corp declared a write-down of $8.4 billion in assets — about 40% of it attributed by Wall Street analysts to the *Journal* deal. Murdoch declared that his empire "may never return to record levels." Did he say "may?" The word is "will." And then there is the sad tale of *The New York Times*—yes, *The New York* Fuckin' *Times*. In early 2009, it was a deer in the headlights of an oncoming cement-truck: a May 2009 $400 million debt payment which it had insufficient cash to satisfy. So first the company announced plan to sell, then lease back, 19 of the 25 floors in its brand-new headquarters. Then it suspended its stock dividend and borrowed $250 million at usurious rates from Mexican oligarch Carlos Slim, whom a *Times* editorial not long ago condemned as a "robber baron." And if not Slim, who—Loanshark.com?

All of this pain is going on, curiously enough, as the overall audience for newspaper content is dramatically expanding. Many readers, young and old alike, are logging on to newspaper web sites, but they don't pay for the privilege. Because they've never had to. Nor will they even put up with advertising to underwrite their news Jones, because they don't like that, either. The unspoken compact between media and consumers—having to endure commercial messages as the *quid pro quo* for free or cheap content—has never applied to Generation Y and will be difficult to *impose ex post facto*. Never mind that the generation's intellectual-property ethos—"All Content Wants to be Free"—is stupid and criminal on the face of it. They truly believe that malarkey, and aren't apt to change their minds.

How about magazines? Rivers of blood there, as well. In 2008, news-stand sales—i.e., the profit engine of the industry—fell 12%. According to the newsletter *MIN*, gross ad pages in the first part of 2009 dropped a staggering 22%—that coming off of a dismal 2008. Conde Nast has folded *Domino*, Meredith has folded *Country Home*, Ziff-Davis has folded *PC Magazine*, Hearst has folded *Cosmo Girl* and *O at Home*, The New York Times has folded *Play*, Hachette has folded *Home*. *Playgirl* is gone. *Radar* is gone. The formerly weekly, formerly bi-weekly *U.S. News* is now a monthly. Time Inc. magazines reduced headcount—mostly by layoffs—by 1400 employees—since 2004. And *TV Guide* magazine, the erstwhile 17 million-circulation goldmine, was sold in October to OpenGate Capital for $1, or $2 less than a copy at the supermarket checkout. At the Magazine Publishers Association conference in May 2008, Time Inc. Chairman-CEO Ann Moore told colleagues, "If you're sitting on your five-year plan, you're delusional."

Down the Tube

Naturally, I save the most shocking old-media victim for last: the extraordinary travails of TV. After nuclear fission — and possibly, though not certainly, the automobile — television has been the most powerful force of the 20th century, our principal form of entertainment and our window to the world. Oh, well. Nothing is forever. Without being overly simplistic or melodramatic, the state of the Old Commercial Broadcasting Model can be summarized as follows: a spiraling vortex of ruin.

According to Nielsen, in the new millennium, the U.S. TV audience eroded an average of 2% a year — even though, in the same period, the population increased by 30 million. According to Swivel, in 2000 Americans devoted an average of 793 hours to broadcast TV and 104 to the internet, a ratio of just under 8:1. By 2008, with broadband penetration in the U.S. tripling, the TV/internet ratio had gone to 675 – 200, or 3.4:1.

The cost to advertisers of reaching 1,000 households with a 30-second spot in prime time, according to Media Dynamics, has jumped from $8.28 in 1986 to $22.65 in 2008 — but effectively more like $32, because between 150 and 200 of those 1000 households use TiVo to skip past the ads. As of 2009, according to Magna Global, 30% of U.S. homes were equipped with TiVo or other digital video recorders. Not only does time-shifting of favorite programs render network schedules irrelevant, between 50% and 70% of DVR users skip past TV commercials. By 2012, DVR penetration is projected to be at least 40%. Cable's long-term prospects are no better. Though it has swallowed much of the networks' audience, it's no more DVR-proof than broadcast. And it is also a victim of a sort distribution auto-immune disease, wherein the body attacks itself. The very co-ax it the industry been stringing for the past

50 years is now the pipe for broadband, which households increasingly are using to bypass pay-cable entirely. As Verizon CMO John Stratton told the 2009 *Ad Age* Digital Conference, "As customers find new ways to acquire content, habits will form, and those habits will be very difficult to break." At which point, he warned: "the business is going to gradually slip out of the back door."

Meanwhile, there is the sword of Damocles called "cost." The reality-TV fad has enabled networks to fill their ever-more-irrelevant schedules, and blindly cast for hits, with cheap programming. But how much longer will they last? Westerns and spy shows, superheroes and hospital dramas all once burned bright. Then they burned out. Yeah, for some inexplicable reason, Americans seem to love *Dancing With the Stars*. But once upon a time, they also loved Vaudeville.

In short: So long boob tube, hello YouTube.

Rolling your eyes, are you? You aren't alone. At the 2007 Bear Stearns Media Conference, CBS CEO Les Moonves protested, "We've been hearing this for years: 'The network is dead. The network is dead.' ... All four networks are going to get CPM increases, in the [pre-season ad-sales market called the] upfront. The business is extremely healthy."

Maybe you think so, too. Maybe you believe that vast structures on which vast societies and vast economies depend do not easily lose their primacy. Perhaps you believe that the TV commercial and magazine spread, and radio spot and newspaper classified, are forever and immutable, like the planets orbiting the sun. Good for you.

Now, say hello to Pluto — the suddenly former planet. Immutable, it turns out, is subject to demotion. But to dramatize the idea of fundamental reordering, we needn't go back 5 billion years to the origin of the solar system. Instead, just think back to approximately the day before yesterday. Remember how they used to talk about "the MTV generation?" It was shorthand for the post-Baby Boomers who couldn't be stimulated unless you basically jammed kaleidoscopes in their eyeballs. They had cut their teeth on the rapid-fire editing and visual noise of music video, so all media were obliged to pick up the pace or lose the attention of an entire generation. And just in case the symbolism escaped you, don't forget the first song that ever played on MTV:

"Video Killed the Radio Star," by the Buggles.

An ironic debut spin, eh? But not as ironic as this: the latest thing the MTV Generation has begun losing interest in is MTV, where ratings have fallen sharply. Short Attention Span Theater has changed venues, and is now housed on YouTube. Online video is killing the video star. Over at MTV Networks, the layoffs began in February 2007. And no huge surprise there. At least, not to me. It was way back in 2005, in *Ad Age*, that I first floated "The Chaos Scenario," predicting that the pillars of the old media would soon come tumbling down. That the MTV pillar had Public Enemy and George Michael and *NSYNC posters plastered all over it, and was deemed the last word in modern television, makes it especially noteworthy — but by no means unique. Since that essay was published:

- In December 2005, Viacom spun off CBS, the so-called Tiffany Network, lest the broadcast business impede growth and depress shareholder value.

- In October 2006, NBC announced a $750 million cost cutback, including 700 jobs and a moratorium on scripted programs in the first hour of prime time. In the spring of 2009, it replaced its entire 10 pm schedule of dramas and comedies with *The Jay Leno Show*. At about the same time, Fox announced plans to cut four hours from its Saturday kiddie schedule, returning two hours to affiliates and selling infomercials on the other two. *Infomercials*, on network TV.

- In November 2006, Clear Channel — the left's favorite boogey-man of democracy-destroying media consolidation — sold to private-equity owners and declared that it wants to unload its TV and small-market radio stations. The sale fetched $35 per share. In the year 2000, the stock sold at $100 per share. The new owners managed to sell their 56 TV stations, but the radio sales were held up in the credit crunch, and with debt default in the offing, S&P in early 2009 rated its corporate paper five grades *below* "junk." The harbinger for all this came in 2003, when stations started adding ad slots to generate cash, and then quickly undercut their own rate cards to unload the inventory. "When you start devaluing your product" says Marci L. Ryvicker, vice president for equity research at Wachovia Capital Markets, "your business model is broken."

■ Nationwide, according Bernstein Research, 2009 TV-station ad revenue is projected to fall 20% to 30%. The market value of stations has also plummeted. According to Ryvicker, broadcast groups whose stocks in 2003 traded at 16–20 times cash flow now sell at a multiple of only 8. For asset sales, the falloff has been even worse: multiples of 18–24 times cash flow now reduced to 6.

■ In 2008, the total U.S. media spend was down — despite the Olympic Games and the most extravagant election campaign in U.S. history. It was the first time overall spending declined in an even-numbered year since 1970. 2009, with its crippling recession, promised to be much worse.

■ For broadcast networks, in the 2008/2009 season steady erosion became a mudslide. According to Nielsen Media Research, CBS: down 2.9% in primetime. ABC: down 9.7% in primetime. NBC: down 14.3% in primetime. Fox: down 17.5% in primetime. In the week before Christmas, 2008, CBS's *The Mentalist* was the top-rated network drama of the week. Yet the odds are quite lopsided that you didn't see it. In leading all other shows, it drew an audience of 10.7 million people, or 3.2% of the U.S. population. Fifty years earlier, despite the star power of a young Clint Eastwood, the western *Rawhide* could manage no better than 17th place among TV dramas. Yet its audience was 11.4 million viewers, then representing 6.5% of the population. The top-rated drama back then was *Gunsmoke*, with 17.4 million viewers. As a percentage of the population, Marshal Dillon and Kitty and Festus outdrew *The Mentalist* by 3-to-1.

How's that for a fusillade of bullet points? Nor does the disruption end there. If DVR does, indeed, reach 40% in 2012 or thereabout, that is exactly the threshold at which 40% of advertisers say they will dramatically reduce their TV buys. I know that's a tough bit of arithmetic to absorb, so I'm going to repeat it: The U.S. penetration of ad-avoiding DVRs will soon reach the point at which 40% of advertisers say they will dramatically reduce their TV buys. If they do as they say, and nothing more, the broadcast TV business is over. Because the advertisers will flee. And, already, slowly but surely, they are doing just that. After years of steady growth in spite of steadily declining audiences, the broadcast upfront market in 2006 was down five percent. Coca-Cola, never a big upfront player, pulled out altogether. So did Johnson & Johnson, which then shifted $250 million online. And that was before the 2008–2009 depression, the dust bowl economy that finally frightened advertisers in a way that the Law of Diminishing Returns mainly had not. They fled in droves. By February, 65% of national advertisers had slashed spending. The automotive and banking sectors, traditionally worth $20 billion to the media economy, all but disappeared from the equation.

Not to put too fine a point on it: Told you so, told you so.

The Long Arm of Economic Law

Mass media, of course, do not exist in a vacuum. They have a perfect symbiotic relationship with mass marketing. Advertising underwrites the content. The content delivers audience. Audiences receive the marketing messages and patronize the advertisers, and so on in what for centuries was an efficient cycle of economic life. But, as I've demonstrated, as the mass-ness that fed the symbiosis disintegrates, each organism loses the sustenance of the other.

"It's a very different kind of world," says Adam Thierer, senior fellow at Progress & Freedom Foundation, and author of *Media Myths: Making Sense of the Debate Over Media Ownership.* "The problem is, the expectations are there to capture that mass audience that long ago disappeared. We are witnessing the gradual death of the business models that thrived in that age of scarcity."

The value of television, like the value of anything, is built upon the economics of scarcity. For decades, the source of highly-produced entertainment was limited to three or four distributors, i.e., the major networks. Cable expanded the options 10-fold, then, with digital cable, 100-fold. Now the internet promises to do so infinitely. Strictly speaking, as a distributor of goods, broadcast's revenue structure should have collapsed long ago. But TV isn't really in the program-distribution business. It's in the audience-selling business, and there the economics of scarcity still stubbornly reign. Because no other medium offers the scale or reach of TV, advertisers have continued to pay more and more per thousand viewers — which is why right up to the recession Les Moonves was commanding higher CPMs, why the upfront market had not yet plummeted and why video advertising on the internet, according to the IAB, amounted in 2007 to a paltry $324 million. On TV, it was $80 billion. But economics will have its due. The law of diminishing returns will eventually prevail. Those who have perennially spent more and more for less and less will finally say "No more" and take their money online (although, as we shall see in the next chapter, not necessarily to online *advertising*).

"I still love and enjoy TV and believe it is very effective for advertisers," says Association of National Advertisers President Bob Liodice. "But we're killing it. We're gradually killing it with cost increases, the level of clutter, the quality of the creative that is out there."

Or, to put it more simply, consider the verdict of Jim Stengel, the recently retired Chief Global Marketing Officer of Procter & Gamble, a man who had a $6 billion marketing budget and no clear idea of where to spend it: "The old model is broken." It is that. Just to recap: fragmentation has decimated audiences, viewers who do watch are skipping commercials, advertisers are therefore fleeing, the revenue for underwriting new content is therefore flat-lining, program quality is therefore suffering (*The Biggest Loser*, Q.E.D.), which will lead to ever more viewer defection, which will lead to ever more advertiser defection, and so on. And here's one more harbinger of doom: increasingly, broadcast networks are bypassing their own affiliates to deliver programming online.

This began in 2006, when CBS, madly trying to cultivate new online distribution channels, put fall premieres of shows like *Smith* and *The New Adventures of Old Christine* on Google Video. NBC used Yahoo to premiere *Heroes* and AOL to offer sneak previews of its *Twenty Good Years* and *Studio 60 on the Sunset Strip*. And the brand-new CW Network celebrated its debut by posting free episodes of *Runaway* and *Everybody Hates Chris* on MSN. The state of the art has since advanced thanks to Hulu, a joint venture of NBC Universal, Fox Entertainment Group and Viacom (Comedy Central. MTV), which delivers current programming free online — notably with limited, but unskippable advertising. The networks first said these were measures to promote the broadcast versions of their shows. But that story never washed. These were experiments in post-broadcasting distribution — experiments the industry would soon come to regret. In March 2009, Jeff Bewkes, CEO of Time Warner, complained to *The New York Times* that TV "went out and did deals to put content on broadband without a whole lot of thought about the long-term financial model." Not that there is one, to speak of. In any event, by that time previous network-executive pronouncements about "audience building" had already been exposed as, um, "inoperative" in December 2008, when the heads of both CBS and NBC publicly acknowledged contemplating life beyond broadcast.

"Do we want to be what we've been?" asked Jeff Zucker, CEO of NBC Universal. This was a rhetorical question — one that boded poorly for those invested in the status quo. Zucker's weak schedule and Incredible Shrinking Audience had already forced him into huge spending cutbacks (spun, a bit unconvincingly, as "NBC 2.0"), yielding cheap, even-less-popular programming (no dramas or sit-coms in the first hour of prime time and, later, *The Jay Leno Show* only in the last hour) leading to still more viewer defection and so on towards oblivion. Thus Zucker's public musing about a previously unthinkable proposition: once affiliate contracts and pro-sports deals expire, his company ceasing to be a network at all. NBC: the cable channel.

The obvious immediate victim would be local affiliates, which get a big chunk of their revenue from selling commercial space within network programs. The internet, needless to say, bypasses them. But here's the thing: it doesn't bypass you. Until about five minutes ago, remember, almost all video-entertainment content was produced and distributed by Hollywood. Period. That time is over. There was a time when advertisers could count on mass audiences for what Hollywood thought we should be watching on TV. That time is all but over. There was a time when broadband penetration was too slight and bandwidth costs too prohibitive for video to be watched online. That time is sooooo over.

As my handy bullet points above amply demonstrate, both print and broadcast — burdened with unwieldy, archaic and crushingly expensive means of distribution — are experiencing the disintegration of the audience critical mass they require to operate profitably. Moreover, they are losing that audience to the infinitely fragmented digital media, which have near-zero distribution costs and are overwhelmingly free of charge to the user. Free is a tough price to compete with. As documented by Woodward and Bernstein, Deep Throat's advice to unraveling Watergate was to "Follow the money." To understand the current predicament, you must follow the no-money. And when you do, you'll have taken The Chaos Scenario one step farther: to a digital landscape in which content is nearly infinite, in which marketing achieves hitherto unimaginable effectiveness, but in which display advertising will no longer dominate.

"I always found Marshall McLuhan annoying," says Bruce M. Owen, senior fellow at Stanford University and author of the seminal *Television Economics*, "but the medium conditions the message. It's already happening."

Denial and Other Bold Action

So what's it like to face your economic mortality? There are some clues in a February 2007 speech by Timothy Balding, CEO of the Paris-based World Association of Newspapers: "What we are seeing completely contradicts the conventional wisdom that newspapers are in terminal decline.... The fashion of predicting the death of newspapers should be exposed for what it is — nothing more than a fashion, based on common assumptions that are belied by the facts." Balding's set of facts comes courtesy of the proliferation of skimpy freebies, like *Metro*, which are to newspapers what Skittles are to cuisine. Meantime, the Swedish daily *Post-och Inrikes Tidningar* dropped its print edition to publish online only. That might not seem too significant, except that the paper had been printed on paper for the previous 362 years. Balding, somehow, didn't mention that development.

Such rosy outlooks, however, are not unusual. For the longest time, even as the structures of their businesses were swaying before their eyes, the captains of the industry professed optimism — even militant optimism. Jack Kliger, then-president-CEO of Hachette Filipacchi Media U.S. and chairman of the Magazine Publishers of America, declared in spring 2007 that "We are no longer threatened by digital media." Perhaps he didn't notice the precipitous drop in readership, what with the industry-wide circulation fraud and all. Or perhaps he was busy killing *ElleGirl* and *Premiere*, but never mind. He was dug in: "I'm not ready to end up my career watching our industry get marginalized and fade away." (Yes he was. The following year, 2008, newsstand sales of magazines — the profitable side of the business — plummeted by 12% and Kliger's French bosses made him a non-executive chairman and started slashing away at costs, including withdrawal from the MPA).

Likewise David Rehr, President and CEO, National Association of Broadcasters, who greeted the National Press Club in October 2006 as follows. "Ten months ago, when I took this position at the NAB, I knew that joining the broadcasting industry would be exciting. But after seeing the dynamics of this business first hand, it is 20 times more exciting than I could have ever imagined." Naturally he's excited. Ratings are plummeting. The networks are bypassing his members via the web. The cash value of stations is in decline. What's more exciting than piloting a plane in a tailspin?

As for Les Moonves of CBS, who bragged about charging his customers more for less, please note that pride goeth before the fall—just as, in Elizabeth Kubler-Ross's stages of death, before bargaining and acceptance goeth denial. (At another media conference only year and a half later, Moonves was more accepting, floating the idea of CBS converting to a cable channel.) But at the Bear Stearns confab, unreality ruled. Protesting that DVR ad-skipping isn't so menacing, Time-Warner's (then-COO) Bewkes trotted out perhaps the most absurd rationalization ever proffered: "When you fast forward, you get a quick visual version that is three seconds instead of 30, you could get the same message anyway."

The :30 is dead! Long live the :3!

How could such a high-priced talent offer such a plainly defensive and idiotic argument? The key words may well be "high" and "priced," if you know what I mean. "It's the cash cow scenario," says Bob Greenberg of the IPG interactive agency RG/A. What CEO will risk his share price, his job and personal compensation package to cut the cow in half? They'd rather just wait, Greenberg says, and hope nothing too disruptive happens too soon. "It's like Bush," he says, "turning the war over to the next president." In the interest of full disclosure, I should acknowledge that the rejection of traditional-media doomsday scenarios is not limited to those with a proprietary stake in the Old Model.

For instance, when asked if he felt like the last nail in broadcast TV's coffin, YouTube's Chad Hurley reacted as if he were facing a space cadet from the Planet Moron. "Why is that?" he asked. Shelly Palmer, of Advanced Media Ventures Group, says, yeah, the consumer is in control ... and the consumer wants to watch TV. "There's no way that changes while we are alive, no matter what anybody thinks. As long as Americans are paid on Friday, then Thursday night from 8 to 11 is going to be the most important time to reach a major audience. It will be incumbent on broadcasters to deliver that audience, and they will, because that's what they do." Then there was that other fellow.

Me: "Do you buy the Chaos Scenario?"
Bill Gates: *"No ... You'll see a little bit more turmoil, in terms of*

> *who succeeds and who doesn't, but it's not some overnight*
> *cataclysm."*
>
> **Me:** *"Is it fair to assume that the advertising people actually*
> *permit into their lives will be more informational and less,*
> *let's just say entertaining and creative and whimsical than*
> *advertising we've seen in the past?"*
>
> **Gates:** *"I wouldn't say that."*

On the other hand, of the $500 million Microsoft allocated to the introduction of its Vista operating system, 30% went online. If every national advertiser did the same tomorrow, Madison Avenue and Hollywood wouldn't be chaotic. They would be Pluto, immediately and forever relegated to some barren, subordinate outer orbit of the economy. And a lot of people would be singing a different tune. Maybe it would be "Katyusha," as rendered in a Montenegrin harbor. Or maybe this:

> *And now we meet in an abandoned studio.*
> *We hear the playback and it seems so long ago*
> *And you remember the jingles used to go.*
>
> *Video killed the radio star.*
> *Video killed the radio star.*
> *In my mind and in my car,*
> *We can't rewind; we've gone too far.*
>
> — The Buggles, 1979

Chapter 2

THE POST-ADVERTISING AGE

*D*OVIÐENJA, BUDVA, MONTENEGRO. *Olá*, São Paulo, Brazil. Here, for most of the past 60 years, advertising hasn't been so much an industry as a cult — a cult of glitz and celebrity, not to mention sex, drugs and rock and roll. What Hollywood is to the U.S.A., advertising is to Brazil. Its practitioners are superstars, household names, gossip and paparazzi fodder. And their output is — along with Ipanema Beach, the Amazonian jungle and *Carnaval* — part of the national identity. If this odd cultural fact were lost on the foreign visitor, it wouldn't take long to figure it out. A visit to São Paulo, Brazil's largest city, was like a journey into an advertising funhouse. By 2006, the city had amassed 13,000 billboards, one flashier than the last. To stand in a high-rise overlooking Avenida São João and Valle Anhangabaú — the confluence of seven streets and avenues into a vast plaza — was to see cityscape as pinball machine, a hurly-burly of images and typefaces, blinking lights and irregular shapes, logos and exposed flesh writ very, very large. Today, the same view yields a tableau of streets and buildings, passing cars and pedestrians, but not a single ad. No semi-nude models, no light shows, no logos — as if an expanse of a square mile had be digitally retouched and scrubbed clean by Naomi Klein.

One of the architects of this transformation stands at a 15-story window and approvingly regards the scene. Matilde da Costa is Director of Projects, Environment and Urban Landscape, for the city of São Paulo. Since 2006, her office has presided over the complete de-billboardification of the city. So inundated had São Paulo become with commercial messages, the administration decided that local color had mutated into blight and, in the space of a few months, under the penalty of heavy fines, it all came down.

"It's marvelous," says da Costa, with a shiver. She's sleeveless in linen and white sandals on a chilly day in the waning days of March summer, so it's hard to tell if she's emotional or just cold. But she certainly seems pleased with herself. "Now I again see the postcard beauty of Avenida São João."

Maybe so, but what about Brazil's self-image as a mecca of advertising creativity? Of all places, how could Sao Paulo dismantle the cult's idols? This wasn't merely removing objects of worship; it was phlebotomizing the life's blood of the national culture! Desecration! Sacrilege! Murder! Eh, no problem. Polls showed 65% of the population was delighted with the whole thing. So much for the public love affair with advertising.

"Now we can see the city," says de Costa was quiet satisfaction. "Before, we couldn't feel. We could only read."

An advertising-free landscape. What a curious idea.

Here's a *Quarterlife*. Call Someone Who Cares

The first element of The Chaos Scenario, as we've seen, creates an inexorable death spiral, in which the fragmentation of audience and DVR ad skipping lead to an exodus of advertisers, leading in turn to an exodus of capital, leading to a decline in the quality of content, leading to further audience defection, leading to further advertiser defection and so on to oblivion. The refugees — audience and marketers alike — flee to the internet. There they encounter the second, and more ominous, Chaos component: the internet's inherent limitations. For all the revolutionary impact it is having on humankind, the web cannot precisely replace the media it is helping to destroy. It comprises more "content" than has ever existed since the dawn of man, but it will not likely be a channel for newspapers of the scope we've come to take for granted, nor for the sort of highly-produced TV programs that have so saturated our culture and shaped our lifestyles for 60 years. At least, not anytime soon.

No, those institutions evolved as a direct manifestation of the mass-media/mass-marketing symbiosis. In an aggregated micro-media/ micro-marketing universe — in the absence of direct payment by consumers — there is no business model to sustain them. Ask Marshall Herskovitz. He's one of Hollywood's most admired producers. His film credits include *Traffic, The Last Samurai, I Am Sam, Blood Diamond* and *Legends of the Fall.* On television, he and partner Ed Zwick created and produced *thirtysomething, My So-Called Life, Once and Again* and, most recently, *Quarterlife.* On that project, a drama about seven young artists in the throes of uncertainty about their lives and careers, he had hoped to have a successful television show, a successful web series and a social network built around the whole synergistic thing. At this writing what he has is a withering social network, ranked 107,403 by Alexa, no TV show and a lot of dearly won cocktail-party conversation.

Oh, there was a powerful coming-of-age drama, all right, but most of it took place in his offices.

"I have proven your theory correct, much to my dismay," he tells me. " 'The Chaos Scenario?' I am living it."

Quarterlife was originally envisioned as an ABC series, but the pilot, by Herskovitz's own admission, just didn't work. After he and the network disagreed over the rewrite, they parted ways. That's when Herskovitz experienced a producer's epiphany. This was the 21st century, after all. YouTube was soon to be invented. What if "you could create and distribute stuff on the internet and own it yourself, and no network, no distributor?" The number of viewers would be smaller, but so would production costs, and many fewer fingers would be in the pie. Furthermore, no network meant no network interference, no "notes," no casting demands, no schedule shuffling and, of course, no cancellation.

But Hollywood is still Hollywood, and established producers simply don't fund experimental programs with their own money. So after a flirtation with Disney (that ended when Herskovitz and Zwick objected to demands for Disney-style cross-promotion), they wound up cutting a deal with MySpace. Though the idea of a social network built around *Quarterlife* represented a potential conflict, MySpace still agreed to host 24 webisodes — and to help fund the four hours of total content — in exchange for half the ad revenue.

Then something else wonderful happened. NBC jumped in, too, script unseen, seeking neither ownership nor control. Just a share of the ad revenue. This is what they call "the best of both worlds." Or so it seemed. Herskovitz called in every chit he'd ever accumulated and shot four hours for a song.

"We premiered on MySpace in November 2007," he says. "We were very, very successful on the internet. Within 6 months, we became the most successful scripted internet series ever. We ended up with 300,000 views per episode."

That is a large number of people in internet terms. It is not a large number of people in advertising terms — especially advertising that is asked to underwrite Hollywood content, no matter how frugally produced. Even the 9 million views it accumulated over three months did not constitute a profitable audience. But, hey, no problem: this was also going to be a TV show, which would draw more people online, which would feed the social-networking aspect of the website, which would create more awareness for the TV show, and so on.

"We were the first web series that graduated into a television series," Herskovitz fondly recalls, "which was very exciting."

Then came the network premiere, February 22, 2008. Despite three months of web previews and a built-in audience, *Quarterlife* the TV show drew but 3.9 million viewers — the worst in-season performance in the network's 10 p.m. slot in at least 17 years. In the demographic most coveted by advertisers, adults 18 – 49, the show had a 1.6 rating. Then, as Herskovitz understates, "All hell broke loose, basically." Basically, the show was cancelled after one episode.

The remaining episodes were immediately dumped, without promotion, on NBC's Bravo cable channel and that was that. The TV life of *Quarterlife* was over — and that's when the synergy kicked in. The webisodes soon ran their course. In the face of the network belly-flop, MySpace wasn't interested in ponying up for more production.

"Without a television partner there was no way to finance the episodes," he says. "The advertising revenue from MySpace was not enough to finance production."

Yeah. That. Herskovitz and Zwick had run headlong into the economics of online media. Even when CPMs are high (let's say the $23 peak network primetime averaged in 2008), and even if you could run three ads with an 8-minute webisode, a show with an audience of 300,000 can generate but about $20,000. That doesn't make the nut in Hollywood. That doesn't pay the bottled-water bill in Hollywood. Still, Herskovitz harbors hope.

"I think this is doable. I think this eminently doable. A million viewers per episode, then it would become a viable business."

Of course, that presumes somebody would bet on the come. Without the entrenched system of TV upfronts, which generate ad guaranties to finance production, producers would be reduced to going door-to-door, hat in hand, as in movie projects. And movie projects sometimes take a decade to finance. Which is why, as the pioneer of online Hollywood ruefully observes, "Nobody's done it yet."

In a post-advertising age, very possibly, nobody ever will.

Supply and . . . Even More Supply

In the good old days of minimum choice and maximum audience, ad revenue generated vast sums of money for a handful of content distributors. In the ultra-fragmented world of the internet, that critical mass of revenue will be available to Google and, for the foreseeable future, nobody else. This would be the case even if online ads offered marketers the same intrinsic value as the traditional-media counterparts. But, for several fairly obvious reasons, they do not. The first is, as mentioned in the previous chapter, that annoying Law of Supply and Demand.

"It couldn't be more straightforward," says Randall Rothenberg, president and CEO of the Internet Advertising Bureau. "Today the average 14-year-old can create a global television network with applications that are built into her laptop. So from a very strict Econ 101 basis, you have the ability to create virtually unlimited supply against what has been historically relatively stable demand."

So big publishers, whether MSN or the *Washington Post* or CNN, with all their vast overhead, have no more access to audience than Courtney the 8th grader. And there are hundreds of millions of Courtneys, millions of them on Google Adsense, driving the price of ad space down, down, down. This was all too apparent in my conversation in January 2009 with Brian Tierney, publisher of the *Philadelphia Inquirer* and *Daily News*, who noted that one third of his 60 million page views a month generated no ad revenue whatsoever. "Clearly a free internet model online — if you build it, they will come — I don't think is working for media like ours.... I think we're going to have to start to find a way to charge for it and not just rely on advertising." Unfortunately for Tierney, he was a day late with the idea, and several hundred million dollars short. Three weeks later, his Philadelphia Media Holdings filed for bankruptcy.

Or think about Yahoo. At about 3.5 billion daily page views, it is the most visited website in the world. In 2008, it had profits of $424 million on $7.2 billion revenue. Not too shabby, unless you compare it to 2005, when the company had profits of $1.9 billion on revenues of $5.3 billion. Last spring, after a prolonged game of chicken, it finally rejected Microsoft's takeover bid at $33 per share. That would have amounted to $50 billion. As of March 2009, Yahoo traded in the range of $12, for a market cap of $17 billion. Microsoft has an outstanding offer for Yahoo's search business, but nothing else. What does it mean when online usage soars, yet the most popular publisher's value is cut by two thirds? It means that Wall Street sees Yahoo's margins headed steadily down — not just because it gets trounced by Google in search, but because its CPMs — the prices it fetches per thousand page views — are going in the wrong direction.

So supply and demand: it's maybe online's biggest problem, but not the only one. Randy Rothenberg is painfully familiar with all of them. In addition to being an old pal and former *Ad Age* colleague, Randy is not only as smart as anyone in the business but disarmingly candid. For instance, he is quick to acknowledge his industry's second boogeyman: ad avoidance, evidenced by average click-through rates below 3%. Drives him up a wall. Yet, for all his brains and economic realism, Randy's constituency is the online-ad industry, and he stubbornly insists there's a solution: "Better advertising. More informational. More entertaining. More beautiful." A latter-day Creative Revolution, that is. Just as Volkswagen and Avis and Alka-Seltzer and Benson & Hedges overthrew the hegemony of the hard sell back in the 1960s, he believes, inspired online advertisers can learn to engage users on the web.

"I'm happy that the interactive industry is finally and belatedly beginning to see that the way we built our sites, based on the direct-response foundation, infused the environment with ugliness and clutter. But direct-response is really the one area in the advertising business that has no institutional concern for aesthetic, regardless of what the long-term effect on the brand is."

Beauty and ingenuity trumping indifference and naked hostility? He's not alone in that fantasy, either. When you ask David Jones, CEO of Euro RSCG, how his agency network is equipped to deal with the atomization of audience, here is what he has to say: "The huge thing for our industry is that actually, what we have been great at for the last fifty years and what we will be great at for the next fifty years is developing and delivering entertaining, engaging, short-format content."

Huh? Content? And all this time we thought advertising is the crap that interrupts content. Yet it's easy to see why Jones might actually believe that bill of goods. When every September at Advertising Week in New York your ad characters draw a parade crowd on Madison Avenue, or when Adriatic tourists pay for a Night of Advertising in the middle of their seaside vacations, of course you feel like you've got a kind of mini-Hollywood going. You might even feel loved. But, sorry, David, you aren't, especially. In fact, your advertising is mainly resented. A 2006 Forrester Research survey found that 63% of respondents believe there are too many ads, and 47% say ads spoil their reading or viewing enjoyment. This isn't just talk. Depending on whose numbers you believe, between 50% and 70% of DVR users skip past ads. The historical *quid pro quo* — acquiescence to advertising in exchange for free or subsidized content — is yet another casualty of the revolution.

"The more access people have to technology," says consultant and social-media entrepreneur Peter Kim, "the more they will use it to skip advertising. When you as a consumer want content, you just want content. You don't want to be interrupted."

Nor is there any reason to think interruption is better tolerated online. On the contrary, Forrester reports that only 2% of consumers trust banner ads, and 81% of broadband users deploy spam filters and pop-up blockers. Have you ever clicked on a banner ad? Ever? Nobody else has, either. Not to sound negative or anything, but all ads are spam.

Hard to see, then, how David Jones' spiffy "content" can much alter those attitudes. Granted, fewer dancing silhouettes and pop-ups and cutesy IQ tests might be nice, but if you need to trump Econ 101 and basic human behavior, better creative might not do the trick. The job calls for something a bit more valuable and elusive. Like a magic beanstalk. Keep in mind that the three most widespread internet phenomena — Facebook, Twitter and the $1.65 billion Google acquisition called YouTube — haven't among them earned enough ad revenue to get a sailor in trouble on a three-day liberty. The laws of economics are not easily violated. In early 2009, Wenda Harris Millard, then co-CEO of Martha Stewart Living, told a conference audience that the glut of supply is depressing prices for publishers both online and off. "Advertising simply cannot support all the media that's out there," she said. And at which conference did she say that? Randy Rothenberg's IAB, where she is chairwoman. Her previous gig? Head of sales at Yahoo.

And the Lord said ... Nothing About "Reach"

Hey, this isn't just me calling the emperor naked. The world's biggest marketer, Procter & Gamble, has been talking about this for 15 years. When CEO Chairman A.G. Lafley says "We need to reinvent the way we market to consumers," he doesn't mean "We need to find a place to amass 30 million people at a time so we can tell them not to squeeze the Charmin." As his CMO Jim Stengel told the American Association of Advertising Agencies media conference in the spring of 2007, "What we really need is a mindset shift, a mindset shift that will make us relevant to today's consumers, a mindset shift from 'telling and selling' to building relationships." And he did not specify media advertising as a way to do so. In fact, his examples — from word-of-mouth to social networking — mainly had nothing to do with advertising. And why should they have?

Mass advertising flourished in the world of mass media because they were — by a happy economic accident — mutually sustaining, not because they were part of God's Natural Order. You've read the Ten Commandments; not one of them is "Thou Shalt Finance Hour-Long Dramas," nor is there a word in there about "reach" or "scale." So why assume that these must transition to the new model? Not only is it economically nonsensical, it squanders the very nature of the digital universe, the ability to speak with — not to, but *with* — the narrowest communities and individuals themselves. Thus the third problem with the future of online advertising: in a connected world, ads are a kind of crude and clumsy means of creating relationships with consumers. After all, people may not much care for commercials, but they like goods and services just fine and are in constant search of information about them. Oddly, in its obsession with not repelling audiences, advertising over the past two decades has provided more and more production spectacle, more and more belly laughs but less and less actual information. Very quickly, because information is at its very core, the online world will soon enough fill the vacuum.

Once again, it is not merely my hunch but history itself demonstrating that display ads will not be the principal means of communication. The annual online-ad spend in 2008 was $23.4 billion. Of that sum, according to Randy's own IAB, 33% was spent on display, while 44% went to search. Why? Because search is contextual, measurable and information-rich. The double-edged sword of search, of course, is that it captures shoppers in the process of shopping, but does little to build brand awareness for the general population. On the other hand, building brand awareness for the general population is wildly inefficient. As online display advertising itself becomes more targeted and measurable, it will be best deployed as a sort of street signage — posted on extremely vertical social networks or served based on user profiles — directing the audience to where the real information is: brand or third-party websites, or embedded in highly utilitarian content.

"I guess the most important thing that I would be asking myself," says media economist Bruce Owen, "is 'How can I make advertising something that people are not only willing to put up with but actually have positive willingness to take?'"

Considering the statistics you've just been reading, that sounds almost preposterously naïve — like asking commuters to vote for traffic. But not only is there an answer to his musing, that answer presages a Golden Age of marketing. The fact is, people care deeply — sometimes perversely — about consumer goods, from Tag Heuer to North Face to Tab. What they don't like is being dictated to about what they should care about or when they should be caring. Forrester's research reveals that 48% of consumers believe it is their right to decide whether or not to receive ad messages. Opt-in emails were deemed twice as trustworthy as TV commercials and 10 times more trustworthy than banner ads precisely because the consumer chooses whether to engage. This may be culturally difficult for advertisers to accept, having spent two centuries trying to browbeat/seduce captive audiences. But take heart. Once the consumer is in the driver's seat, he or she will often cheerfully drive right in your direction.

"I'm amazed that anyone would go online to the American Express site to learn about credit cards," says Ted Shergalis, founder and Chief Product Officer of [x+1], a web optimization firm. "but they do. By the millions every day." Yet the same consumers may TiVo right past Ellen DeGeneres in an Amex TV spot, because, "They want the information on their terms."

No, brand advertising will not disappear. It will, however, be less ubiquitous, and its very nature may well change. As *Wired* editor-in-chief and Long Tail proponent Chris Anderson likes to observe, "Brands are a proxy for information." In other words, brandedness itself conveys to consumers a minimum assurance of quality, reliability and distribution. Obviously, brands convey other things as well — things like values, status and personality. But branding's most basic function will be usurped by the information readily available at a mouse click, or an instant PDA scan at retail. One particularly eye-opening application, from Vancouver, Canada, is VideoClix, a hypervideo tool that lets the user roll over any part of the image — a car in the background, for instance — and click for information about make, model and so on. A second click directs the user to the manufacturer, retailer or whatever. It's like VH1's old "Pop-Up Video" show, only the user alone controls what to pop up. Thus it exploits the online third dimension, beyond audio and video: info-depth. "It's a layer of information," says founder Babak Maghfourian, "that people will demand."

Or consider Nike Plus, the joint project of Nike and Apple in which the iPod becomes a tool for monitoring your running pace and style, and for fitting a custom pair of shoes. The Nike Plus website combines utility, community, information and, of course, online sales. It is the marketing program, the CRM engine and the store. The sole function of the TV commercial — which is an elaborate demo notably devoid of celebrities and narrative and jokes — is to drive traffic to the site.

Now, multiply that formula times the Leading National Advertisers plus the entire Long Tail of trailing advertisers. What I'm describing is the democratization of the information economy, simultaneously destroying fortunes and creating them. When order is someday restored, surviving marketers will be in clover. Others will be entirely left behind.

To some of those "others," the reality is finally beginning to sink in. When the yin is shriveling up and dying, after all, this should hardly be lost on the yang. For instance, listen carefully to Jan Leth, executive creative director of OgilvyInteractive North America, as he tells a funny little story about an agency assignment for Six Flags.

"They had a promotion for their 45th anniversary. They wanted to give away 45,000 tickets for opening day to drive traffic. So we got a brief to do whatever: ads, microsite, whatever. But our interactive creative director just went off and posted it on Craigslist. Five hours later, 45,000 tickets were spoken for. No photo shoot. No after-shoot drinks at Shutters," he adds, with faux regret. Then, with somewhat less irony: "Now, the trick is, how do you get paid?"

That cute little anecdote represents something between grim news and total doom for the advertising-agency business, which continues its erratic Pluto-like orbit around marketing budgets as if unaware that it has lost its stature. Circumstances have conspired to threaten its place on the cosmic map altogether. Ad agencies are simply not organized in a way to profit from modern means of connecting with consumers. So they pay lip service to the digital future while digging their heels into the 30-second-spot present. In case you skipped past this book's introduction, listen to this from a June *Times* of London op-ed by Sir Martin Sorrell, chairman of WPP Group, the world's largest marketing-agency holding company:

"Slowly, the new media will cease to be thought of as new media; they will simply be additional channels of communication. And like all media that were once new media but are now just media, they'll earn a well-deserved place in the media repertoire, perhaps through reverse takeovers — but will almost certainly displace none."

No need to panic, he says. The internet's just a new channel, like gas-pump video or checkout-aisle coupons. It's just a question of managing the transition. Yeah, sure it is. If Sir Martin were being honest with his readers, he'd admit that in terms of culture, organization, expertise and compensation structures, a global ad agency can no more easily transition from a Gross Ratings Points mentality to a world of aggregation, information, optimization and Customer Relations Management than Young & Rubicam can transition from English to French. It's two entirely different grammars and vocabularies. Not to mention revenue models. The Brave New World, when it fully emerges, stands to be far better for marketers than the old one — but not much thanks to display advertising. As we shall see in the next eight chapters, they won't need much of that to connect with consumers. Which is why most existing ad agencies and some media agencies will be left behind. And the reason they will be left behind is their stubborn notion that they can smoothly transition to a digital landscape.

An ominous episode occurred in 2007 when Nike — the multi-billion-dollar brand that emerged and dominated thanks to the advertising genius of agency Wieden & Kennedy — decided that, when it comes to fulfilling all of its digital marketing needs, Wieden just couldn't do it. *Wieden & Kennedy!* It's probably the greatest independent agency in the world, but, by Nike's lights, too invested in old thinking to suit the brand it almost single-handedly built.

Note, too, that while Sir Martin is publicly spewing happy talk, he's also been reconfiguring WPP's portfolio so that less than half of the business — on the way to only 1/3 — derives from advertising or media services. No wonder. Agencies make money making spots and ads, and buying the media for them. Furthermore, in order to exploit the internet's phenomenal capacity for targeting and optimizing messages in ads and on websites, advertisers will have to invest vast resources in information technology infrastructure — hardware, software and flesh-and-bloodware — to crunch the vast amount of data that will be pouring in every second of every day. In the aggregate, this will amount to many billions of dollars. Much, if not most, of the money will come from existing ad budgets. That cannibalization will only further accelerate the destruction of mass media/mass advertising symbiosis and unlock the very power of aggregation, information, optimization and Customer Relations Management that will render most image advertising impotent and superficial.

In other words, Madison Avenue has problems out the yin-yang.

So Where Do We Get New Episodes of *The Mentalist*?

So, in the post-advertising universe, who pays for content? Ah, *the* question. One reason it is so difficult to imagine The Chaos Scenario is the sense of entitlement we have come to harbor about free access to the finest (or, at least, most expensive) output Hollywood has to offer. The sub-prime crisis takes away my house, we've got a problem. The Chaos Scenario takes away my *House*, and we've got a revolution. Alas, I can offer no especially reassuring answer. The best thing I can say is that the world's gravitation online to YouTube, et al, already reflects mutations in our entertainment DNA. Obviously, the internet alone offers more content — instantaneous and free — than mankind has ever had at its disposal in all of history. Though 99% of it is unwatchable, unlistenable, unreadable *dreck*, the 1% still represents an astonishing cornucopia of brilliance — including some material far cooler than anything television has ever had to offer. Aggregation websites such as Fark, Boing-Boing, ebaumsworld, CollegeHumor and Digg, and the sheer filtering efficiency of social networks, do a pretty good job of refining the gold from the ore. Others actually combine the functions of aggregation and social-networking. Magnify.net hosts thousands of vertical video-sharing communities and Ning.com thousands of equally vertical social networks — from "American Idol Fans" to "Asthma Parents."

Marc Andreessen, founder of Netscape, is co-founder of Ning.com. "People are interested in what they are interested in," he says. "The magical part of social networking is the people [specific category] advertisers are interested in are magically coming together." Other sites, such as

Will Ferrell's FunnyOrDie.com, combine professional and user generated content to cobble together something like a cumulative mass audience — and the theoretical possibility of ratcheting up the revenue to sitcom levels. Still, as Marshall Herskovitz can sadly attest, that's the longest of long-shots. But what if in the near future most content is paid for by the user, either via subscription, like HBO, or a la carte, like pay-per-view or iTunes? This would eliminate advertising from the equation. If micropayments ever become practical, pay-as-you-go would allow users to seamlessly buy, for instance, newspaper content on an edition-by-edition or even article-by-article basis. As media economist Bruce Own puts it, 'The willingness to pay by consumers is far greater per eyeball than the willingness of advertisers."

In 2008, citing the pioneering efforts of Wal-Mart, Amazon.com and iTunes, Adams Media Research projected that paid streams and downloads will quickly overtake advertising as the revenue model for video content. "By 2011," according to the report, "advertiser spending on internet video streams to PCs and TVs will approach $1.7 billion, but movie and TV downloads will generate consumer spending of $4.1 billion." Another 2008 report titled "The Digital Consumer: Examining Trends in Digital Media," this one from the investment firm Oppenheimer & Co., concluded the same: content "is not likely to be ad-supported."

Such a revenue model would suit content producers, as well. In their life-and-death struggle against piracy, no digital-rights-management technology offers the awesome impact of price. As download costs are pushed inexorably downward — say $1.99 compared to $25 for a DVD — the incentive to steal is pushed down accordingly. Why bother shoplifting from the dollar store?

Chapter 3

THE WORLD'S MOST SUCCESSFUL FAILURE?

L OOK, BEFORE YOU EVEN GET TO THE second paragraph, try this: Go to YouTube.com. In the search field, type in "boom goes the dynamite." A video thumbnail will pop up. Click on it and watch.

It's just a little outtake from a Ball State University campus TV newscast. It features a courageous but overmatched freshman named Brian Collins presenting the worst sports-highlight rundown in human history, culminating in the worst sportscaster catchphrase ever conceived. "Boom goes the dynamite." The video is horrifying. It is cruel. It is hilarious.

Search around some more. Type in "evolution of dance," which has gotten north of 115 million views. You wouldn't think "Ohio motivational speaker's grand finale" would equal "mesmerizing," but Judson Laippley's seamless sampling of footwork to 30 songs, from Elvis to 'NSYNC, pretty much is. Or try the accurately titled "Noah takes a photo of himself everyday [sic] for six years." A time-lapse documentary of some guy named Noah Kalina over 2,356 days, it's a little thin on plot, but it's nonetheless racked up more than 12 million views. Which is nothing compared to the latest YouTube sensation, Susan Boyle, who amassed 80 million for various clips over the course of a single April 2009 week, thanks to her magnificent voice but mainly to her touching ability to sing while ugly. And, on the subject of the revulsion/attraction paradox, you'd better also see "Numa Numa," which stars a chubby young man in his New Jersey bedroom lip-synching to an insipid but weirdly fetching Romanian pop song. Or, what the hell, live dangerously. Type in "sweet tired cat" and watch a drowsy kitten dozing off. The clip, which was viewed nearly 2 million times in two weeks, is 27 seconds of such concentrated cuteness that you might actually have a stroke and die. It's that excruciatingly adorable.

And, as it turns out, extremely valuable. In 2006, Google paid $1.65 billion in stock to become sweet tired cat's home.

The price tag for YouTube, just to put the investment in perspective, is what Target paid for 257 Mervyns department stores and four distribution centers in 13 states, and just a bit more than what WPP Group paid for the Grey Global Group advertising network (10,500 employees in 83 countries generating $1.3 billion in revenue). Mervyn's and Grey, of course, were at the time both profitable enterprises with vast fixed assets. YouTube's fixed assets pretty much consisted of a video interface, some server farms and a cool retro logo. So why was it worth nearly six times the gross domestic product of Micronesia? And, beyond money, what is its greater significance? What does the YouTube phenomenon tell us about the Listenomics Age? What will YouTube *become?*

This chapter will definitively answer those questions.

Well, maybe not exactly answer.

But explore.

OK, speculate. But it's an exercise worth undertaking, because it gets to the very heart of The Chaos Scenario and Listenomics and involves nothing less than cultural, sociological, and economic transformation — including, but not limited to, a reallocation of the $80 billion that advertisers spent on TV in the U.S. in 2008. That upheaval would require a couple of things to fall into place: 1) a business model to convert what is basically an overgrown fan site into a profitable advertising medium, and 2) a tectonic change in the worldwide media economy. But, as we've seen from the last two chapters, Item Two is well underway.

As for Item One, well, it doesn't look especially promising, but don't be too quick to bet against Google. Not long ago, all *it* had was a search algorithm and a cool logo. Now, after reinventing online advertising, it has revenue of $22 billion a year and good reason to believe that neither of those daunting prerequisites is out of the question. In January 2009, according to comScore Video Metrix, 101 million people viewed 6.3 billion YouTube videos, and on average spent six hours doing it. In the U.S. alone. Six hours! Up from zero hours in January 1999. Obama schmobama. *That* is change.

"Just as our kids don't understand the difference between broadcast and cable," says Buzzmachine.com blogger Jeff Jarvis, "the line between TV and internet TV is about to disappear."

Jarvis calls the phenomenon "exploding TV," and YouTube is exploding faster than anything else: from a standing start about a year ago to more than 200 million video-streams a day. It was on YouTube, not *Saturday Night Live,* that the world fell in love with two nerdy white guys rapping about their "Lazy Sunday." It was there that we found ourselves smitten, intrigued, and ultimately betrayed by Lonelygirl15. And it is there that more than 15 hours of video are uploaded every minute, their creators posting what they think are video clips but that are also improvised explosive devices laying waste to the old order. Hit "Upload Video"...

Boom goes the dynamite.

Monkeyvision

Chad Hurley says he doesn't remember. I spoke to him back in 2006, exactly two weeks before the announcement of the Google acquisition, and he'd just flown the red-eye to New York to make his case to Madison Avenue. He'd be turning right around in a few hours; but at the moment he was stuck in yet one more conference room, and his eyes had the vacant look of someone whose body has a one-bar wireless connection to his nervous system. In a word, the man was fried. Never mind that he was the cofounder of the Next Big Thing and was poised to be a total tycoon; the question on the floor seemed to have him stumped: What was the first video uploaded to YouTube by someone other than himself or YouTube cofounder Steve Chen?

He insisted he can't quite recall, you know, the $1.65 billion moment.

"I think it was a few people from Stanford," Hurley offered. "People in a dorm room doing weird things."

Weird things? What kind of weird things, Chad?

"I don't even remember," he said. "That was so long ago."

Yeah. Way back when: 18 months earlier, in May 2005. But you can scarcely blame the young fellow if it's all a bit of a blur. In the intervening year and a half, YouTube hosted millions upon millions of clips — all because Hurley and Chen wanted a utility like Flickr for sharing videos. In a garage in Menlo Park, California, they built a simple interface and a one-click way to embed videos on other sites. It was a serendipitous innovation, coinciding with the MySpace phenomenon and yielding what Hurley calls "this" — as in, "It's all turned into this." Slouched wearily in his stackable conference chair, he looked a little bewildered, but maybe it's more like bemused. Ain't like he has no explanation for all this. You'll find it in the company slogan: "Broadcast Yourself."

"Everyone, in the back of his mind, wants to be a star," Hurley asserted for probably the quadrillionth time, "and we provide the audience to make it happen."

Lots of people can now watch themselves on sort-of TV, which is pretty fun in itself. The bonus is that others want to watch them, too. Third-millennium humanity has demonstrated an interest in sifting through millions of pieces of crap produced by total strangers to discover a few gems — some accidentally entertaining ("Boom Goes the Dynamite"), some breakout performances from the previously obscure ("Treadmill Dance"), and some explorations of a new art form crackling with genius (Ze Frank, Ask a Ninja, and the guys behind Loneygirl15.)

Throw in the uploaded TV commercials, such as Dove's "Evolution" or Nike's Ronaldinho spot showing the Brazilian soccer star miraculously, repeatedly volleying against the crossbar. Add to that some professional content either stolen from or surrendered by Hollywood. Altogether, this stuff constitutes a bottomless reservoir of short-form video content for others to siphon off if they choose. Which they do, millions of times a day, from pages all over the internet. It's said that if you put a million monkeys at a million typewriters, eventually you will get the works of William Shakespeare. When you put together a million humans, a million camcorders, and a million computers, what you get is YouTube. Inventing Monkeyvision, however, is cold comfort if you can't build an actual business around it. Which is why, on the eve of the Google deal, Hurley went to New York: to explain to advertisers why they should give him money to broadcast *them*selves. The bad news for his entourage that day was that advertisers have been broadcasting themselves for decades and would honestly prefer the status quo. The good news, as previously explained, is that the status quo isn't long for this world.

Still, even when the big deal was announced, the financial reckoning of YouTube's value hinged on a series of what-ifs.

What if there were a means to approximate the reach and mesmerizing power of television online? What if there were a medium with not only the grip of TV but the vast scale to absorb all those ad dollars? And what if, as a bonus, the medium were able not merely to command eyeballs for marketers but to target content especially relevant to what the marketer is selling? In short, what if there were a missing link between the old model and the glittering new one? What would happen then? Actually, that's an easy one: Procter & Gamble would be ecstatic. Blood would flow in the gutters of Madison Avenue and Hollywood. And Ryan Seacrest would be out of our lives forever.

Oh, and someone would strike it filthy, stinking rich — and not the piddling few hundred million Hurley and Chen each earned, either. We're talking real money. Some of it would be dispersed along the long tail of video-sharing Web sites: Fox Interactive, Yahoo, Hulu, MSN, CBS, AOL, Viacom, Turner Network, Disney Online, NBC, Break

Media, ABC, DailyMotion, Glam Media, Metacafe, ESPN, Megavideo, Photobucket, CollegeHumor, Blip, Ning, Heavy, Revver, FunnyOrDie, DailyRadar and on and on and on. But, as of mid 2009, YouTube had 43% of the market. I'll just say that again: *43% of the market.* As Cervantes put it: Holy Fuckin' Toledo!

Constituo, Ergo Sum

I'm busy, really busy, trying to write this damned book. But I'm also compulsive, and I toggle back and forth to my email constantly, just in case a lucrative speaking gig has come in, or I've been laid off, or an attractive investment opportunity has cropped up in Nigeria. Just scanning the subject line, is all; I don't actually open anything because I don't have time ... unless.

Unless it's from one of my daughters, with a YouTube link.

That I open, because I know my 3-minute time detour will be rewarded with something absolutely delicious — like a little girl officiating, pre-flush, at a bathroom funeral for the late goldfish, Lucky. Furthermore, I know I'll have the subsequent delight of laughing about it with them, comparing notes on our reactions, reliving the experience and generally hooting our asses off. That's just part of how, and why, YouTube has altered human behavior on a global scale. As Chad Hurley puts it, "Everyone wants to see what everyone else is seeing and enjoying." Which is nothing new whatsoever. On the contrary, it is one of the last remnants of mass culture. What Uncle Miltie and the Super Bowl and *Survivor* have always offered is something to talk about at the water cooler, at the nail salon, or on IM. "Did you see that horrible sportscast? 'Boom goes the dynamite!' What is that supposed to mean? ... Wait. You haven't seen it? Ohhhhmygosh! I'll email you the link."

"It goes back to something primal," says Henry Jenkins, director of the Comparative Media Studies program at MIT and the author of *Convergence Culture*: *Where Old and New Media Collide.* "There's still a desire to have a shared cultural context. We hunger for things we can discuss."

But why has anybody bothered to upload the clip in the first place? Whether it's video of them lip-synching a Bucharest chartbuster, or Lucky's farewell, or a montage of TV-news bloopers or a pirated excerpt from *Saturday Night Live*, what do they get out of it? Well, it's what Hurley says: the opportunity to broadcast yourself—whether your literal self or your inner self, reflected in what kind of toilet-flushing-dog home videos most amuse you. As the fashion, cigarette, liquor and bumper-sticker industries have known for a long timer, it is impossible to overestimate man's innate desire to announce to others Who I Am. Remember *Death of a Salesman?* Remember poor Mrs. Loman, trying to explain to her sons the essential truth about their tragically insignificant dad, Willy? "Attention must be paid," she said. She understood everything. YouTube is the revenge of Willy Loman. Or, belatedly, his salvation. It quenches a desire never before quenched in all of human history: the hitherto futile aspirations of Everyman to break out of his lonely, anonymous life of quiet desperation, to step in front of the whole world and *be somebody.* An Accenture study of 1,600 Americans found that 38 percent of respondents wanted to create or share content online. Thus, suddenly, the inexplicable "Numa Numa" begins to make sense. Why would a portly New Jerseyite film himself being ridiculous? Because he'd be on (sort-of) TV, on his own terms. This is the heart of the thing.

"If you aren't posting, you don't exist," says Rishad Tobaccowala, CEO of Denuo, a new media consultancy. "People say, 'I post, therefore I am.'"

Constituo, ergo sum. An interesting formulation that may well represent a new rationalism for the digital age. But for the moment, let's not put Descartes before the horse. Because all of this amounts to nothing if it should disappear as suddenly as it arrived.

A Zillion Eyes. Now Monetize!

Admittedly, there's a slight possibility I was mistaken about that Cervantes attribution, but this I'm sure of: somebody definitely once said 100 million people can't be wrong. Obviously, YouTube (like its successor killer apps Facebook and Twitter) is immeasurably valuable to users. But is it valuable in dollar-and-cents terms? Is it an ad medium? Is it a business? Four years into bewitching the entire world, YouTube has so far found the answer: not even close. It turns out that success is 1 percent inspiration, 99 percent monetization. Listen to John Montgomery is Chief Operating Officer of GroupM Interaction, a media-buying arm of the WPP Group communications conglomerate. "They've got the audience. In order now to monetize what they've got, they need to figure out a revenue model. But it's a very, very hard thing to do around user-generated media."

He told me that, by the way, in early 2007, shortly after Google bought in. Montgomery certainly was prescient. Monetizing popularity, he understood, is no easy row to hoe. According to Bear Stearns, YouTube's 2008 ad revenue was $90 million, which might seem like a chunk of change if it covered, say, the expense of copyright-infringement litigation, which it probably did not, or bandwidth costs, which it certainly did not. Credit Suisse estimates operating costs north of $700 million. That's because, almost three years after its big acquisition, Google has yet to lick the two gigantic structural problems of the online world. As discussed in the last chapter, 1) the glut of online ad inventory reduces the price fetched by any given publisher to chump change. 2) nobody wants to look at ads, period.

YouTube long refused even to sell ads appended to either end of a video — like a TV commercial — on the grounds of safeguarding the viewer experience. To Montgomery, users regard them not as the fair price for viewing content but as media authoritarianism, "as punishment for us supplying content you've got to see this ad" — which, he hastens to add, happens to describe the broadcast television model. Which TiVo and YouTube are destroying.

No wonder, then, that many accept as an article of faith that a pre-video commercial is fatally intrusive. "There may be a user out there who likes a 30-second pre-roll," says Tod Sacerdoti of POSTroller. "If you find one, have him give me a call." According to four years of data accumulated by video network Vidsense, 15-second pre-roll ad causes

8% of the audience to abandon the clip before it starts. A 30-second pre-roll sends 22% of the audience packing. As for Sacerdoti's so-called post-roll ads, even the most self-satisfied marketer wants to know who in the world would stick around to watch. "Nobody will watch post-roll," Montgomery says, and he's not exaggerating by much, Vidsense claims that its data indicate 76% of viewers flee before the end the ad. Certainly you've experienced that very impatience and disgust as you've sat through a pre-roll, or, more likely, not sat through one. You just feel abused. In fact, in the world of YouTube, people felt abused before the ads ever showed up. The mere fact of the Google deal, back in 2006, spawned bitterness and indignation at the grassroots. YouTubers lamented the inevitable advertiser plundering of their pristine environment, a cherished oasis of relative noncommercialism in a world of brand inundation. Many were dubious that 'GooTube' would retain its soul. "I think its the beginning of the end of youtube as we know it," wrote a poster named SamHill24. Another, Link420, declared simply, "ITS OVER!!!! youtube is screwed."

So how did Google solve the problem? Until late 2008, by *not* solving it — putting off a day of reckoning while first so dominating the video marketplace as to crush any and all competitors aborning. Winning market share by selling at a loss, in other words. If it were Japanese steel Google were flooding the market with, instead of kitten videos, it would be called dumping. "Google, like Facebook, is playing a game of chicken," says Troy Young, chief marketing officer of Videoegg.com, "The game is, 'Until we find a better way to monetize, we're going to bias our philosophy towards user growth, destroy our competitors, and at some point in the future re-evaluate where we're at.'"

Beware of Dead Cats

As to the first part, check: 43% market share, after all. That astonishing domination, and the astonishing change in consumer viewing behavior it suggests, certainly emboldened advertisers to rush to YouTube ... albeit not with their dollars. Rather thousands and thousands of times they tried to exploit the vast potential audience at no cost whatsoever, by uploading entertaining ads as videos and letting virality just take its course. YouTube actually encouraged this, for what it was worth — and that wasn't much. For one thing, as WPP's Montgomery was quick to observe, for advertisers "the most successful way of using YouTube" — posting ads for free — "is a way in which YouTube doesn't make any money." YouTube did give it the ol' college try, telling prospects that buying ads to complement those free uploads stimulated the viral effect, a dubious claim that most of Madison Avenue simply ignored. The other problem was that — as we shall see in very next chapter — "the most successful way" of using YouTube was almost never successful. Virality isn't a strategy; it's a fluke. In the history of online video, you can count the viral smashes on the fingers and toes of one amputee. This has changed the advertiser calculus of "success" in ways bordering on the absurd. In March 2009, an ad-agency spokesman characterized a McDonald's spot as "a YouTube sensation." In four days it had generated 50,000 views. Or about what you'd get with a single :30 on *Good Morning Seattle*. Three weeks later, another advertiser sent out a press release claiming its video "has been dubbed a success, after being viewed on YouTube alone 40,000+ times in the first 5 days of the campaign." (You couldn't blame them for not understanding the internet. It was some outfit called "Microsoft.")

In the matter of viewer ad tolerance, YouTube did have one thing going for it: there's no audience backlash to an ad if there's no ad to begin with. For the first four years of its existence, very few YouTube clips — fewer than 1% — were deemed eligible for ad content of *any* kind, either because the clips included unlicensed intellectual property, such as a Bono song or an MTV outtake, or because advertisers couldn't be sure the videos represented a positive (or even neutral) environment. Given the anything-goes nature of the web, advertisers are in constant, paralyzing fear of inadvertently associating themselves with violence, pornography, hate speech, or God knows what lurks out there one click away. "Advertisers and brands are enormously risk averse,"

Magnify.net's Rosenbaum says. "The question now is how the raw and risky is made safe and comfortable. It's not a little question. It's a big question." For instance, if you are, say, Meow Mix, and you bought ads adjacent to cat-related videos, how surprised and disappointed you might be to learn you have sponsored a YouTube video uploaded by someone named mrwheatley and titled "exploding cat." Or the one from qu1rk89 titled "exploding cat." Or this one: "ma907h eats dead cat," which shows a guy ... oh, never mind. As media consultant Cory Treffiletti grimly observes of the metadata describing a YouTube video, "Right now it consists of only a few key terms the user selects. And there's no blank to fill out for 'cat vivisection.'"

Man on the Moon

To Vidsense CEO Jaffer Ali, whose sites consist of 75% licensed "legacy" content for the precise reason of being advertiser friendly, suitability questions alone doom YouTube to commercial failure: "the inherent danger to a brand of going on YouTube by being next to the guy lighting his farts on fire." But that, he says, isn't even YouTube's fundamental weakness. Nor the crippling bandwidth costs. Nor the pitiful ad rates in a supply-glutted marketplace. No, to Ali the issue is elemental: the online ads don't work. The pre-rolls are too brief to serve as brand-image builders, and the click-through rates of .3% make them all but useless as direct-response vehicles. People snacking on video bits are simply not in a transactional mood. "It's proven not to work," says Ali, who came to online marketing from the world of TV direct-response. The key to *that* business model was the combination of low ad rates for marginal programming in bad time slots, and — he insists — the very marginality of the programming itself. "That's why direct-response always worked on the shit that nobody wanted to watch, because, 'If I miss it [to dial the 800 number for the advertised product], I'm not missing anything.'" In other words, the very pistachio-nut quality of watching YouTube videos reduces to near zero the chances any viewer will follow any ad to another destination.

In short, he says, "Forget it. They're never gonna make any money with YouTube."

That is what you might term the pessimistic view. The optimistic view is that, as a sustainable advertising vehicle, YouTube has much to overcome. But that doesn't mean Google is giving up. This is America. We have a fighting spirit. There's no such thing as a problem, only an opportunity. It's always darkest before the dawn. When the going gets tough, the tough get going. Or, as President John F. Kennedy said, in his speech promising a manned mission to the moon, "Why does Rice play Texas? Not because it is easy, but because it is hard."

The question is, assuming Troy Young's conclusions about a two-pronged, build-now/cash-in-later strategy are correct, has Prong Two commenced in earnest? Chris Allen, vice president and director of video information at the media-buying firm Starcom, believes it has. "I think they've got to a point where they said, 'Ohmygosh, it's do or die time. We've got to start monetizing this thing.'"

True enough, as of the spring of 2009, YouTube was offering advertisers a wide array of options — each of them, in its own way, useful and each, in its own way, problematic. One employed technology developed by Videoegg to superimpose "overlay" ads, stripped along the bottom 20% of the viewer. Overlays quite resemble those obnoxious bottom-of-the-screen promos on Fox and Bravo — one of the reasons you're not watching TV in the first place. At the moment, these overlays overlay a small fraction of videos, but even that sliver takes a toll. Vidsense found that the presence of overlays on the sites in its network reduced page views 17%.

Not long after offering that option, YouTube also realized that if the real-estate adjacent to the video-player is most monetizable, why not sell more of it? So instead of ordinary banners, it now sells mega-banners. These are the entire width of the screen at the top of the page, with an option to go down the right rail. Essentially, they're rich-media billboards, a virtual Sunset Boulevard — especially since they are mainly from entertainment companies, who are the one category so far to significantly embrace YouTube. And why? Because YouTubers apparently are predisposed, by golly, to enjoyin' a movin' picture show. So much for the ultra-targeting ability of the internet.

Of course, YouTube is gingerly trotting out ultra-targeting, too. In March 2009, the company introduced so-called "interest-based" ads as follows:

> *These ads will associate categories of interest — say sports, gardening, cars, pets — with your browser, based on the types of sites you visit and the pages you view. We may then use those interest categories to show you more relevant text and display ads.*

> *We believe there is real value to seeing ads about the things that interest you. If, for example, you love adventure travel and therefore visit adventure travel sites, Google could show you more ads for activities like hiking trips to Patagonia or African safaris. While interest-based advertising can infer your interest in adventure travel from the websites you visit, you can also choose your favorite categories, or tell us which categories you don't want to see ads for. Interest-based advertising also helps advertisers tailor ads for you based on your previous interactions with them, such as visits to their websites. So if you visit an online sports store, you may later be shown ads on other websites offering you a discount on running shoes during that store's upcoming sale.*

If that all sound strangely like behavioral targeting, that's because it *is* behavioral targeting, a means of keeping track of people's movements on the internet in order to divine which advertising they might deem more relevant. (discussed in more detail in Chapter 7, "Guess.") Ads with greater relevance can command 10 times greater a price than one served randomly. The targeting technology, however, has become radioactive since a controversy in 2008 involving the company Nebuad, which partnered with internet service providers to track the online patterns of subscribers, in some cases with insufficient notice and essentially invisible opt-out option. Never mind that Microsoft, for example, tracks your every keystroke via your Explorer toolbar, and that such captured data can be connected only to an IP address, not an individual.

Nor that Google itself collects and stores your every search. This is stuff that makes privacy alarmists foam at the mouth, and next thing you knew Congress was involved. It also didn't help that Nebuad was essentially a clone of the notorious spyware company Claria, perpetrators of countless unsolicited pop-ups for a decade.

"The word some people at Google didn't like was 'targeting,'" says Shishir Mehrotra, whose title is long but straightforward: director of project management, YouTube monetization. Why? Because targeting sounds sinister, evoking images of mean, mean men with rifles and telescopic sights. "But, yes, 'interest-based,' that's the same thing."

Comically euphemistic as it may be, YouTube's "interest-based" ad program is pretty much as the company describes it, with a clear-cut benefit not just to the advertiser (less wastage) and the publisher (more revenue) but to the consumer — namely, if you're going to be bombarded with advertising, better to be bombarded with advertising that has at least a remote chance of mattering to you. On the other hand, as YouTube viewing tends to be eclectic — even random — certainly compared to the vertical media so heavily influenced by personal affinities. That's apt to throw targeting algorithms for a loop. The user has just viewed "sweet tired cat," "Numa Numa," "The Evolution of Dance," "Boom Goes the Dynamite" and 15 Carrie Underwood videos. Precisely which ad does YouTube deliver?

For some users that use the platform infrequently," Mehrotra acknowledges, "it's very difficult to assign interests to them." Still, he says, there are ways to divine individual (IP address) interests. "One of my favorite examples is finding soccer lovers. They watch some soccer content but they often watch other things. But people who are not soccer lovers watch no soccer content."

Okay, that's something — albeit nothing compared to the advantages of tapping into the viewer's Google *search* history, extremely rich data that would enable precise targeting. Alas, for political reasons, YouTube dare not even speak such things aloud. Therefore, for now, as an advertising platform, it is handicapped by the very eclecticism that makes it so popular.

"YouTube is a schlock house," says David Wadler, CEO of Twist-age, creator of customizable online video platforms. "It's like going to the Dollar Store. 'Hey, a $1 plunger!' " And targeting cheap-plunger-purchasers won't much drive up the price of the ads.

So, just to review, Google's YouTube fortunes hinge on:

1) viral ads that produce no revenue
2) annoying overlays that a prominent ad network insists are "proven not to work"
3) pre-roll ads that send 22% of the audience running in the opposite direction and irritate the rest
4) huge billboards advertising movies to people watching videos on computers
5) the Russian-roulette risks of appending ads to sketchy user-generated content, rendering (as of 2009) 91% of inventory unsellable
6) tracking/targeting technology that has already triggered the ire of the House Committee on Energy and Technology and is therefore undeployable.

And then there is a seventh little issue: as mentioned earlier, the troublesome tendency of Joe Laptop to upload video content wholly or partially stolen from its rightful owners.

Every now and then you read about someone who is sorting through a table of odds and ends at a flea market and comes across Aunt Sadie's necklace — stolen in a break-in three months earlier, along with other family heirlooms. That's how Viacom CEO Sumner Redstone feels watching *The Daily Show* excerpts on YouTube. Nobody from the YouTube corporate organization copied and uploaded Viacom's copyrighted material — that was done by dozens or hundreds of elves in the vast YouTube-o-sphere. But YouTube hosted the material and let the rest of the YouTube-o-sphere views the stuff for free. Which is why the ink was no sooner dry on the Google acquisition (whereupon YouTube's pockets went from stitched shut to vastly deep) that Viacom sued for

$1 billion, claiming its property had been streamed off of YouTube 150 billion times. The suit is still alive, and for good reason. Whether sitcom excerpts or a cat flushing the toilet to "Innagaddadavida," copyrighted material cannot be exploited commercially — except under narrow "fair use" exceptions — without explicit license from the copyright owners. In fact, under the Digital Millennium Copyright Act, it's unlawful to even host it once the owners inform YouTube that it's there. Shortly after lawsuits were filed by Viacom and the English pro-soccer Premier League, YouTube installed Content ID technology that compares new uploads against a 100,000-hour master cache of copyrighted material for the purposes of sniffing out problematic videos. That, however, hasn't relieved Madison Avenue's *spilkes.* As eager as advertisers are to push stacks of chips online, in the back of their mind is Napster. It, too, was a peer-to-peer revolutionary — one killed aborning by copyright infringement issues. Nobody wants to invest only to see the fledgling industry paralyzed with litigation, regulation, or legislation. And it is not an idle fear.

This Ain't No Clip Joint

Let us, however, give these people the benefit of the doubt. It is, after all, Google we're talking about — a company that turned a search engine into the most lucrative online money machine in the world. These are not clueless people. For instance, all of the above assumes that YouTube remains an aggregation of home video, semi-pro video and very little else. But, as of the spring of 2009, even as it was expanded its palette of ad options, Google made clear the big second strategic prong has little to do with kitty-cat clips and sportscaster outtakes at all. Rather, it announced licensing deals with Hollywood studios and CBS for long-form content — TV shows and full-length feature films — to be supported by pre-rolls, mid-rolls, post-rolls and probably Kaiser rolls and egg rolls. Like Hulu.com and Jaffer Ali's Vidsense, YouTube's licensed content will be a safe harbor for even the most skittish marketer.

"They're basically joining the bandwagon of becoming a distributor of syndicated content," says Starcom's Chris Allen. Thereupon, presumably, to subsidize all that is unique and mesmerizing about YouTube with the revenue from just another new Hollywood distribution channel. That's a monetization strategy. It could work. And, some believe, must work. We've already heard from John Montgomery, COO, of MindShare Interaction, the huge digital-media-buying arm of Martin Sorrell's gargantuan holding company, WPP Group. Now let's meet John's boss, the CEO, Rob Norman.

"My belief fundamentally is that push media is *not* dead," Norman says. "Unless it becomes a robust display advertising channel then there is no YouTube, and robust means users who — like it or nor — are forced through an advertising experience" — even if they dislike it, resent it, feel violated by it, and regard it as a betrayal of Google's corporate watchword, "Don't be Evil." "You know what?" Norman proposes. "Tough shit. Nobody likes it. It is what it is. The value exchange has to move backwards."

Backwards, he means, to a time when we all took our ad medicine and shut up about it. Here is a man clearly frustrated with the destruction wrought by the very digital onslaught he has made a career of understanding. "It's kind of like the inverse neutron bomb, isn't it?" And indeed it is — a weapon of mass-media destruction that leaves the people alive but topples every structure. But to Norman, the solution is not to meekly surrender to the evolving patterns of digital-media life. This is a matter of life and death. YouTube has no choice but to invade, to impose an ad-supported model by sheer force: pre-rolls, post-rolls and mid-rolls, the whole arsenal.

"You can go public with users," Norman says, "but you stay public with revenue."

That's true, and YouTube's new aggressiveness could indeed force users to rethink the value equation and their willingness to suffer commercials online. Furthermore, if Mehrotra has reckoned correctly, opt-in ads such as YouTube "promoted videos" and behaviorally-targeted ads will not only command higher rates, but will drive less relevant, less engaging ads from the space. "And that," he says, "will eventually drive up CPMs" still farther.

Maybe. But even if they do, in terms of long-term profitability, YouTube, the business, is not necessarily saved. For one, as we saw in Chapter 1, to siphon off audience from Hollywood and the networks is to push *those* industries ever closer to the brink. Once the TV model is in rubble no fresh content will exist for YouTube to serve. Chris Allen calls it a recipe for a future "shockwave." Put another way: a classic zero-sum game. The other risk is that the licensing fees and exponentially greater bandwidth required to be a prime distribution of long-form video will quickly suck up all the additional ad revenue.

These are the sort of obstacles that make you wonder if, in 2006, entrepreneurial geniuses or no, Google's eyes were bigger than its stomach. If their valuation metrics were influenced by the stock-market bubble, they may well have fallen prey to what former Fed Chairman Alan Greenspan famously described as "irrational exuberance." MindShare's Rob Norman, for one, is confounded by the market psychology behind the strategies of YouTube, Facebook, Twitter and other killer apps with *corpus delecti* financials. Not only have they put audience ahead of revenue, he says, they've actually avoided getting income at all, lest Wall Street start demanding larger multiples.

"If the numerator is zero it's kind of hard to multiply," Norman wryly observes, as one in a series of hyperbolic punch lines that get to the bizarre truth of the matter. "God forbid they should send out an invoice. It would be a strategic train-wreck."

The numbers do evoke a sort of '90s déjà vu. Could these Google guys have anted up nearly six times the GDP of Micronesia because they were afflicted *with* micronesia, a small case of memory loss about, say, the insane multiples squandered on fiber capacity just before the telecom crash? Eerily enough, $1.65 billion is just what Racal Telecom fetched from Global Crossing. Ah, yes, the titans of yore. GeoCities. Prodigy. Netscape. Could YouTube be just another interim killer app, a flash in the internet pan, a Numa Numa Neandarthal languishing on Darwin's death row, a historic missing link destined to go missing itself?

Of course it could.

After all, if we must get used to the idea that newspapers will disappear, even though we need and cherish them, and that broadcast TV will disappear, even though we've always considered it a birthright, why should we think that YouTube is safe just because we've quickly come to so love it and depend on it? As the collapsing old media so tragically demonstrate, an audience does not — in a digital world — a market make. At least, not an advertising market. Yet somehow the notion of YouTube failing for lack of revenue seems so ... unthinkable.

Listen to Rishad Tobaccowala, Mr. I-Post-Therefore-I-Am: "What YouTube has going for it is its sheer size. In a fragmented world, there is a need for community and a need for massness."

A need. *A need.* First of all, could the irony be any thicker? The old model is a flaming ruin, disintegrating into nothingness, and what rises from the ashes — in the vast, distributed, exploded, long-tailed micro-media galaxy of the internet — is a mass medium? A general-interest destination? YouTube as the new boob tube? That may not be Jeff Jarvis' idea of the glittering future, but it certainly is Chad Hurley's. As he put it to me, not only without bombast but more or less listlessly as the latest interrogation wound down, "We think people want an entertainment destination."

Yes, when the rest of the entertainment infrastructure is in ruins, we'll all need one. Remember what Henry Jenkins says about shared culture. It's something primal, he says, and YouTube — as much as any other institution of the Brave New World — is well positioned to provide it. But ask the Tribune Company whether the undeniable need for the *Los Angeles Times* and *Chicago Tribune* and WGN helped it meet its debt payments. You'll probably have to go to the bankruptcy trustee to get your answer.

I said earlier not to bet against Google, but if that means counting on advertising sales to achieve profitability — in terms of things that are not easy but are hard — relatively speaking, a wager on Rice vs. Texas is looking better and better.

Oh, YouTube can meet our needs all right. And very possibly it will. But it cannot meet Google's needs until it leverages that 43% market share and starts asking for YouTuBeginPayingYourWay.

TALK IS CHEAP

How quick thy wagging tongue,
Quicker my ear to credit its professions.
Thence with speed the news to market flies
Where fishwives freely trade their new possessions

— William Shakespeare

O KAY, NOT REALLY. Shakespeare wrote no such thing. I pretty much just made that up. But I'm certain he would have weighed in on the subject if he'd had time, because — let's face it — word of mouth is right up his human-nature alley. Everybody knows its power — to spread a rumor, to propel the Next Big Thing, to fill the Globe Theater or to empty it.

Or the multiplex, for that matter. Even without explicit guidance from the Bard, Hollywood studios and distributors know the score. They spend $3 billion a year in the U.S. advertising new releases. But they spend hardly anything after a film's first weekend — because by then they know the matter is out of their hands. If a movie is deemed by audiences to be good, box office will be big. If a movie is deemed bad, box office will be pitiful. And, either way, it is the verdict of those opening-weekend patrons that will prevail. Clever trailers and heavily redacted film-review excerpts ("It's mind-blowing to imagine this trash ever saw the light of day" becomes "Mind-blowing!") may lure people to see an unknown quantity, but ads are useless to influence — much less overcome — negative buzz. Because movie goers are like bees, buzzing back to the hive to exclaim, "Dude, that sucked!" Just ask the producers of *Waterworld, Heaven's Gate, Ishtar* and *Gigli.* On the other hand, the producers of *My Big Fat Greek Wedding, The Blair Witch Project* and *Little Miss Sunshine* can be most grateful for the same phenomenon. Wagging tongues have served them well.

Beyond Hollywood, there are, of course, many other fabled examples of word of mouth yielding enormous results: Cabbage Patch Kids. The spiritual self-help guide titled *The Purpose Driven Life,* Avon's Skin-So-Soft as mosquito repellant — remarkable commercial successes all, with little or no advertising behind them. Or, for a better example still, think about the tire story. You've probably heard it. In the '70s, a woman showed up at a Nordstrom department store in Anchorage, Alaska, to return a tire. This presented two immediate concerns to the sales clerk. 1) She had no receipt. 2) Nordstrom doesn't sell tires. But, famously, the clerk gave her a refund anyway. Why? Well, the transaction cost the store about $30. That investment has yielded about a squijillion dollars in consumer goodwill, because it has come to stand for Nordstrom's legendary attentiveness — a legend that has been cultivated entirely via word of mouth.

"Nordstrom publicly doesn't talk about customer service," says Robert Spector, author of *The Nordstrom Way.* "They don't mention it in their advertising. They will very seldom do interviews about it. It's all done by making it part of the culture. Yet everybody knows about it not only in the U.S. but all over the world." (What they don't know, Spector says, is that the tire story isn't quite as bizarrely beneficent as it sounds. The lady with the tire had indeed bought it in that very building from a store called Northern Commercial, which had recently been purchased by Nordstrom. The clerk made a liberal decision in offering the refund, but not a random one. Yet it still reflected a sales culture short on arbitrary "policy" and long on customer satisfaction, thus resonating powerfully in the consumer psyche.)

Mind you, these examples all occurred long before the digital age, long before email and web pages and blogs and social networks. The news was spread in the most analog fashion: words spilling one at a time out of actual mouths. It can happen with retail chains. It can happen with entertainment. It can happen with a single individual.

The Original Influencer

Consider the case of J.O.N., an incredibly charismatic guy who the WOM mavens call an "influencer." He was a bit of a drifter, and a bit of an oddball, but when he spoke, people listened. Two of the people who listened — we'll call them Pauly and Pete — not only were in his thrall, they talked him up obsessively to everybody, often asserting some rather extravagant claims. You know the urban legends about Pop Rocks and stolen kidneys? They're nothing compared to the eye-openers whispered about J.O.N., but the improbable yarns somehow captured people's imaginations and were passed from one listener to another to another. Even his death, an infamously violent one, did nothing to diminish his popularity. On the contrary, it only burnished his reputation and validated his worldview. In time, even the most bizarre and supernatural details of his legend were not only spread far and wide, but accepted at face value. In Armenia, for instance, where the guy had never set foot, he was an object of worship. Then all over southern Europe. Then Ethiopia. Then northern Europe.

That Jesus of Nazareth. His word-of-mouth was simply outstanding. With no advertising budget whatsoever, his brand soon swept the globe. Today it has 33% market share and 2.1 billion customers. Build a better God, and the world will beat a path to your door.

So don't let anyone forward you a clip of Ronaldhino volleying off the crossbar a few times tell you "viral" is a new phenomenon. BudTV? Puh-lease. The King of Beers was 2000 years behind the King of Kings.

Fine then; it's settled. Word of mouth is a priceless means for spreading good news (or, as we shall soon see, bad news.) But that's obvious, isn't it? The problem for marketers, and artists, and politicians, and anybody else with something to sell is to actually, intentionally stimulate WOM — which, historically, has been about like trying to control the weather. These things aren't manufactured; through an unmanageable convergence of variables, they simply happen. At least, until now. Things have changed. The internet is a word of mouth engine. Sure, binary code is what makes it all work, but the fuel of the digital age is the instinct to share information, whether in social networks like Facebook and MySpace, blogs, Craigslist and Angie's List, Twitter, Digg or YouTube. Even Google. What is the Google algorithm, after all, but the crediting of special value to other people's choices? When you type in a search term, the results you see are determined not just by textual relevance to what you've asked for, but also by how many others have linked to those pages. If you enter, for instance, "moist towelette," you are not first directed to the blog called "Moist Towelette," nor to the many retailers and distributors of moist towelettes, nor even to moist-towelettemuseum.com. No, the first result is the homepage of Modern Moist Towelette Collecting, featuring, among other treasures, the moist-towelette-collectors' anthem:

> *You're Soft*
> *You're Wet*
> *You Smell So Good…*
>
> *Chorus:*
> *I Love You Moist Towelettes*
> *I Love You Moist Towelettes*
> *I Love You Moist Towelettes*

Yes, it's lovely. But dab that tear from your eye and think about why this particular content should get the top position. Simple. There is an active subculture that spends a lot of time there, plus a universe of curiosity seekers such as myself who will drop everything to see what would constitute a WetNap theme song. This is the sort of website that, once it shows up on Digg, gets dug. By their own private choices to click on and link to Modern Moist Towelette Collecting, thousands of individuals are endorsing the site to the rest of the world. The fact that this is accomplished with a keyboard and mouse makes it no less word of mouth. If Jesus had a website, and everyone else had Google, Peter and Paul would have had to do a lot less schlepping.

Song Sung Blue

OK, I grant you it may be a bit counterintuitive — if not actually blasphemous — to juxtapose Jesus Christ with Modern Moist Towelette Collecting. But I stand by my point. Both exemplify how human beings flock to other human beings they identify with and trust. And that poignant trait is of inestimable importance. The collapse of the traditional mass-media/mass-marketing model is being accompanied by digital mechanisms for fluid communication among consumers, who trust one another far more than they trust any banner ad, celebrity endorsement or 30-second spot. According to a landmark 2004 Yankelovich study, 65% of consumers give serious credit to WOM testimonials — versus 27% who say they are influenced by advertising.

Among those with a dim view of marketers dictating self-interested messages, unsurprisingly enough, is Andy Sernovitz, founding CEO of the Word of Mouth Marketing Association, and author of *Word of Mouth Marketing: How Smart Companies Get People Talking.* "Look at Starbucks," he says. "This is a brand that has essentially never advertised and they've built a global brand based on people saying, 'Jeez, that was a really nice experience'" And, once again, that was before social networks allowed human interaction on an unprecedented scale. In the age of Facebook and Blogspot, Sernovitz can't imagine anyone anymore being influenced by advertising, beyond the basic function of the news alert, for a brand new good or service. "It used to be," Sernovitz says, "a manufacturer would launch a product that was fine — not good or bad, but just fine. You spent $20 million developing it, and as long as it didn't blow up, you could sell it for years. And that's how all the average products sitting on our shelves got that way. The new reality is, you spend $20 million developing a product, the day you deliver it it will get reviewed on tens of thousands of blogs and message boards and website. And the success of your product will be determined that day." To wit:

iPhone: good.

Segway HT: bad.

Those verdicts came immediately in. Thumbs up for one, thumbs down for another. The tyrant — i.e., the crowd — was not to be stalled, not to be overruled and, most importantly, not to be seduced by self-interested marketing messages from the manufacturers. Consider the low-fare airline called Song. Amid a whole lot of hype, and a huge investment from struggling parent Delta Airlines, Song was launched in 2003 as not merely a company, but as a fully formed culture. With the high-priced advice of designer and branding guru Andy Spade, corporate-identity firm Landor Associates and web development agency The Media Kitchen, Song packaged an all-encompassing image of look, sound, feel, smell (!) and above all attitude. To them, "Song" ceased to be a noun and became an adjective — as in "That is so Song." Here's what Spade told the producers of a PBS *Frontline* documentary called "The Persuaders:"

I think what differentiates Song is the emotional component of it. If you look at other airlines, especially in North America today, I don't see anyone doing anything that really has resonance on an emotional level that really makes me feel something beyond what's logical or what's practical. You can obviously look at an airline and say, "OK, you have more legroom than they do." And on one level I think that's great, but it doesn't stick with me. The next month I'll see someone else who has another two inches of legroom. So overall you're competing for, in my opinion, a small benefit.

It was an interesting concept, the ultimate exercise in trying to manage the customer's experience purely via atmosphere. Song lasted exactly three years and 15 days. This has something to do with the difference between genuine culture and contrivance, something to do with financial extremis at the parent company and plenty to do with what people told one another about Song. Here's an excerpt of a web review from a customer/blogger named Adam Lasnik:

> *My plane was an hour late getting out of the gate for reasons I don't recall. We were actually all sitting on the plane for that hour, and since this was a red-eye flight, I figured no sweat, I'll just catch some shut-eye in the meantime.*
>
> *Ah, but no, that'd be too logical. The stupid crew had other ideas, blaring really lousy alternative music crap through the main speakers during the entire waiting period. More specifically, they were playing stuff from a new album by The Wallflowers, with whom they apparently have some dumb distribution deal or something like that. How do I know? In their promos, they bragged how Song wasn't just an airline, but also a record label. Oh joy. Just what I want. In an industry where airlines can't even manage their flight schedules or other core aspects of their business, I want to see lots of footage of their execs and such hobnobbing and doing recording deals with lame music artists.*
>
> *Well, fine, I'll just request a pillow from the crew, maybe two, to press up against my head so the music's less loud and I can sleep. Ah, no pillows on board, even for a red-eye (I was told that they're too expensive to clean and store).*

For this guy, the experience had little to do with the flight attendant uniforms, the logo, the atmosphere, no matter how cunningly and methodically planned. He just wanted to get from Point A to Point B, comfortably, on the cheap. No extra credit for a well-thought-out corporate ethos. "Song," Sernovitz says, "was a perfectly executed marketing case study. The uniforms, the theme, the execution was as good as marketing could ever get. But in the end it was fake; it was still Delta. When they walked down the aisle they hand you a CD, everyone on the plane is thinking, 'What's wrong with you? I just want a sandwich.' They blew all their money trying to create a fake user-friendly brand."

Think you can engineer people's attitudes about you? That is so Song.

And Now, Ladies and Gentlemen, Tonight's Top 10 List

And thus can millions — or billions — of dollars worth of paid messaging amount to so much wasted breath. And that is exactly what, amid digital revolution, is taking place. But, of course, you knew that. Nobody disagrees about the power of word of mouth. On the contrary, many a would-be conversation piece lay awake at night scheming about getting the tiger by the tail. The trick is how, and on that point there is hardly any agreement whatsoever. There are those who believe that WOM by its very nature is a spontaneous organic phenomenon entirely rooted in human volition and therefore not subject to third-party influence. Others believe that virtual crowds can be rallied, provided you can identify and impress certain key nodes, or hubs, who have a preternatural influence on those in their social networks. This is the premise behind Malcolm Gladwell's *The Tipping Point*. Still others believe that there is no need to isolate "influencer" demigods, that rigorous seeding of valuable information or content among likely audiences will inevitably sprout far and wide. In this chapter, we'll consider all of the above. Plus, you know, other stuff.

I once read on the internet that one of the best ways to get read on the internet is to do Top 10 lists (if you Google "Top 10" you get 233 million hits. If you Google "trenchant analysis," you get squat). So, in the spirit of trying to get this chapter passed along from one breathless reader to the other, what follows are my Top 10 principles of word of mouth:

1) Listen to the conversation
2) Better yet, host the conversation.
3) Offer that community a stake in your enterprise.
4) Practice jujitsu.
5) Sneeze in public.
6) Have a story to tell.
7) What's in it for us? Not you. Us.
8) Behave yourself.
9) Remember Siegfried and Roy. Especially Roy.
10) Pray for Serenity

As author and copyright owner, I hereby waive my intellectual-property rights with respect to the list and authorize you to photocopy or scan the list and post it prominently above your credenza, laminate it for your wallet or attach to your refrigerator with a souvenir magnet. (Note to FBI: Please look the other way.) So now that you have the Top 10, let's look at them one at a time — some only very briefly because they're explored in depth in subsequent chapters.

Listen to the Conversation

Even if you have no intention whatsoever to try to out-apostle Apostle Paul, and even if you do nothing else to exploit the vast resource of the internet, it is absolutely essential to monitor what is being said on websites, blogs, social networks, Twitter and everywhere else on the online universe. Even without harnessing WOM to spread the word about your brand, by tracking it across the breadth of the internet, you can glean the richest, most bountiful harvest of market intelligence you've ever imagined.

"This is the world's largest focus group," says Jerry Needel, senior vice president of Nielsen BuzzMetrics, the data mining company that provides clients, well, metrics on buzz.

By characterizing things as he does, Needel actually sells his product short. Data mined from online conversations is vastly better than the best focus group ever convened. For one thing, you don't have to serve snacks. More importantly, the opportunity to harvest genuine insights about your brand or your candidate or your policies is several orders of magnitude more valuable than twelve morons behind a one-way glass in the bowels of a shopping mall. Furthermore, the volume of the information is high enough to view it as actionable data — versus focus-group blatherings, which not only don't represent data, they don't even necessarily represent what the subjects actually think. Small-group dynamics being what they are, what is pronounced by those subjects is easily influenced by the moderator or alpha focuser among them. Blog posts and reader comments, obviously, are far less subject to such contamination. Thus, from the accumulation of buzz, you can discover things about yourself and your audience that likely would never reveal themselves in the bunker of corporate headquarters.

In January 2004, ConAgra launched its Life Choice line of low-carb meals, an utterly unsurprising development. The Atkins Diet craze was at its zenith — one in 11 Americans reportedly were trading starch for meat — and food processors were feeling the squeeze. The headline in the *Omaha World-Herald*: "ConAgra Climbs on Low-Carb Bandwagon with New Frozen Food Line."

The thing about bandwagons is that they are all about peak excitement, something which by definition is destined to wane. Initial sales were strong for Life Choice, but BuzzMetrics began to notice something. Skepticism about the Atkins diet began to grow, and overall discussion of the low-carb miracle began to diminish, in chat rooms and blogs online. Six months later, sales of low-carb products, Life Choice among them, took a nosedive. But ConAgra wasn't as vulnerable as most. Whereas Atkins Nutritionals Inc. wound up filing for bankruptcy, ConAgra saw the conversational trend as a predictor and gently backed away from its low-carb play before the bottom fell out of the market.

In marketing, the term of art for the online noise is "buzz." In the intelligence community, they call it chatter, and it's hard to overstate the risk of ignoring it. Recall that there was plenty of chatter in early September of 2001 about an impending terror strike. It went unheeded.

The trick, of course, is separating critical intelligence from meaningless noise. There is a science to chatter crunching, and it will surprise you not at all where much of this science is based. That would be Israel, where failing to connect the dots, on any given day, could spell the end of the republic. There Mossad and the Israel Defense Forces employ some of the country's best mathematical minds to create the algorithms for gathering, categorizing and, most importantly, analyzing billions of and billions of words to divine, for example, the intentions of Hezbollah. There is a great deal riding on these guys knowing what they're doing.

But they also don't stay in the Army forever, and when they leave, many of them turn up in Herzliya Pituach, a high-tech district on the fringes of Tel Aviv often described as the Palo Alto of Israel. It is here, hard by Microsoft's headquarters and just around the corner from Cisco and Motorola, that BuzzMetrics does the math, formulating the algorithms for searching, classifying and interpreting internet content. This is, unsurprisingly enough, a complicated enterprise, taking into consideration such factors as keywords, relationships among keywords, word proximity, patterns and, most significantly of all, sentiment. Gauging sentiment is a particularly nettlesome challenge, because although bloggers and Twitterers and chat room chatters aren't under the influence of some hired-gun moderator or focus-group blabbermouth, it is still hard for a computer program to parse what they're thinking and feeling based solely on word choice.

One reason is that, even to the human eye, characterizing sentiment in text is a thorny matter. Experimentation has shown that given the same text, two individuals asked to categorize words and phrases as positive or negative will agree only 57% of the time. More vexing still is the fact that individuals don't necessarily understand their own sentiments. Our actual behavior often belies not only our assertions, but our beliefs about our beliefs. And, finally, many words are positive in one context, negative in another. "Funny" is a swell endorsement of a sitcom, but not so good to describe the taste of milk.

Enter, then, Yakir Krichman: The "Sentiment Master."

The Master is a lean, boyish 37 years old, who when I met him was sporting a brown skateboarding t-shirt and a day's growth of beard. Both a licensed clinical psychologist and mathematician, he is also working on a PhD in semantic psychology. He isn't exactly a steely sort, but it is hard to imagine the Sentiment Master as *verklempt*. (When I show him a YouTube video of a kitten yawning, the thinnest of smiles crosses his lips. His office bears no photos of his wife and children. Granted, Krichman does keep family snapshots on his mobile, but let's just say that, from all appearances, he is the master of sentiment; sentiment is not the master of him.) Rather, as befits a shrink, he is extremely analytical. For instance, as he tells his marriage-counseling clients, "The opposite of love is not hate. It is indifference. Apathy."

An interesting insight for anyone trying to gain, and keep, the attention of the public.

BuzzMetrics has many clients, but one especially salient example of sentiment-evaluation technology — and a foreshadowing of the so-called "semantic web," in which computers will understand and converse with other computers in essentially human language — was an early experience with HBO. (This in 2002, when the Israelis were a startup called Trendum, which later was to acquire BuzzMetrics, thence to be acquired by VNU's Nielsen and merged with NetRatings to take its current form) They were asked to monitor web traffic to analyze the first season of HBO's *The Wire*. At the time, the Sentiment Master was 31 and his boss, Ori Levy, was all of 25. In New York before their clients, they explained which of the show's characters were most popular, which less. They informed the execs there were some issues with not only the urban jargon in the dialogue, but the particular Baltimore flavor of urban jargon. The main message, thankfully, was that the show would be a huge success.

"I'm 25," Levy says. "I'm standing in from of 10, 15 people at HBO telling them about a show I've never seen."

Oh, yeah, that. None of this was divined from Trendum's analysis of *The Wire* itself. It was all from the online chatter about *The Wire*. And, oh ... the reason they're telling me the story is that their extrapolation was spot-on correct.

Not to be a rainmaker for BuzzMetrics, or anything. There are many free online tools for harvesting the chatter about you — minus, perhaps the granular information, sorting and analysis, but nonetheless chock full of information about who is saying what about you where. I myself use Google Blog Search, Technorati and BuzzMetrics' own BlogPulse about 70 times a day, just to make sure people out there continue to regard me as a sainted genius. Such companies as Comcast — which, as we shall see shortly, generates plenty of chatter, most of it horrible — has a whole team sweeping the net for what we can politely describe as "market intelligence." To get a notion of how busy they are, just Google the term "fucking Comcast." Make some popcorn. You'll be reading for a while.

Better Yet, Host the Conversation

In Chapters 6 and 8, I'll talk at some length about Dell Corp., which at one time turned a deaf ear to the nattering of the *hoi polloi*, but which now embraces the crowd with the zeal of the converted. The scales fell from corporate eyes after management bungled a customer-service complaint from an unhappy customer. The customer turned out to be a highly prominent blogger, who in one bitter post consigned the (then) largest computer seller in the world to an inner circle of hell. "Dell Hell," as it came to be known, because any Google search for the brand turned up "Dell Lies, Dell Sucks." Having more than learned its lesson, Dell now hosts one of the more robust corporate sites — IdeaStorm — where it entertains complaints and product suggestions, and provides a gathering place for customers and technogeeks. In the company's first year as a listener, it had more than 100 million of what it calls "interactive engagements" with the public. Mind you, with a small-scale TV campaign, you can purchase 100 million consumer exposures quite easily. But, there is a vast difference between exposure and a relationship. As many a pervert well understands, when you simply expose yourself, your coveted audience mainly turns away in disgust.

Offer that Community a Stake in Your Enterprise

As I talked about in the introduction — and as you shall soon see in Chapter 8 ("Sometimes You Just Gotta Lego") — it is one thing to have an interest in an organization and quite another to feel a part of that organization. It's not just a sense of belonging, powerful as that emotion is; it's also a matter of pride. Human beings naturally feel more connected, more involved, more motivated and more responsible when they have a stake in the enterprise. Proprietors burn midnight oil; wage earners flee at the whistle. Mortgagees paint trim; renters soil carpets. And nobody doesn't want public education to be adequately funded, but every lady at the school bake sale, I promise you, has a kid in the school. The psychology is so basic, rooted in the potent combination of self-interest and self-esteem, and it can be all the difference between having customers (or voters, or constituents, or audience members) and a community of fellow travelers carrying your flag wherever they go.

The quintessential case study is indeed Lego, where adult fans of the marginal MindStorms line of robotic toys were brought into corporate headquarters (in Denmark, at their own expense) to help plan a second product generation. After months working as designers, they returned home and became active evangelists — yes, apostles — for the company and the second-generation toy. In fact, today — though their official affiliation with the company ended at the end of their volunteer design efforts — their personal websites and blogs constitute virtually the entire marketing program for MindStorms.

As Andy Sernovitz points out, this was Microsoft's strategy for introducing its Vista operating system. "Several million users beta tested the software," he says. "One of the core word of mouth strategies is building community. Sometimes community-building is: let a bunch of people try it and make them feel important and special." Microsoft went on to sell 20 million copies of the new system in the first month — even though Vista sucks (and eventually became victimized by the very community it had so painstakingly cultivated. More on this presently.)

Practice Jujitsu

National Express is the U.K.'s largest long-haul bus company. It's cheap compared to train and air travel, but it has a poor reputation for service, and a consequent problem retaining customers. Management wished to improve the company's culture of service, but also to disarm legitimate grievances before they found their way into public. So they hired a company called Fizzback. Now, on every National Express window is a sticker inviting riders with a complaint — or a compliment — to text, toll free, a special number. The texts are scanned by a computer equipped with artificial intelligence, which responds immediately and asks if the rider would like a call from customer service. Meantime, the text is forwarded to the live human being responsible for the bus route. And whether the problem is a stopped up toilet, or a surly driver, or a broken seat recliner, the appropriate parties are put on the case.

When the bus arrives at its depot, someone from the company is there ready to take care of the problem. In this way, not only do irritated passengers never get the chance to be irate customers, they are often so delighted at the company's attentiveness that they become vocal advocates of National Express. Furthermore, the simplicity of the process encourages satisfied customers to weigh in, too. "More than 50% of the feedback that comes in is positive," says John Coldecutt, Fizzback's director of marketing, "so it's really motivation and creates sort of a virtuous cycle of a service culture." Both the iron hand and the silk glove, in other words, have encouraged frontline employees to be more courteous and diligent, resulting in still better feedback, and so on.

The results, as measured by subsequent research, have been striking. Coldecutt says that National Express's "net promoter score," a standard measure of customer satisfaction, jumps an average of 52% among customers who avail themselves of the free text — whether they are calling to praise the service or give someone an earful.

That's jujitsu: turning the energy of the aggressor into your own best weapon. The Dell nightmare becomes the Nordstrom dream: customers buzzing to friends, family, co-workers and total strangers about their positive experience. As Nielsen BuzzMetrics' Pete Blackshaw likes to say, "Customer service is the new media department."

Sneeze in Public

How nice to think we can introduce a new product, publish a book, launch a candidacy, post a video on YouTube and that sit back and let virulence do its job. Of course, many things would be nice, including, but not limited to: the end of all war, immortality, a water engine and Ashlee Simpson's retirement. Unfortunately, while word of mouth is by definition contagious, it's not all that contagious. It isn't the Ebola virus. It's more like the common cold, spread somewhat inconsistently person to person by atomized gobs of.....

Yecch, this is getting a little gross; maybe "infection" isn't the best metaphor. But there's no shortage of others. Think instead, if you wish, in terms of priming the pump, igniting tinder, jump-starting the engine or, most commonly, scattering seeds. Not randomly or willy-nilly, mind you, because that isn't seeding; it's sampling. Here's the difference:

Maybe you've been in a Whole Foods and stopped by some lady in an apron with a tray full of toothpick-impaled cheese cubes. "Vasterbotten!" she says. "It's Swedish! It's dense, perforated and has a strong, assertive flavor!" You taste one and — sure enough — its flavor is both strong and so charmingly assertive that you buy a cylinder, instead of the Havarti you usually get. That is successful sampling. But when you get home, you don't drop the shopping bags on the kitchen floor, dial a friend and shriek: "You gotta try Vasterbotten!" In fact, you've probably forgotten about the purchase altogether. Exponential transmission requires more than mere discovery, or even satisfaction; it requires enthusiasm, and it behooves the marketer to do its sampling among people apt to be enthused. Mass marketing is on death's door, after all, because it is so irretrievably inefficient. The cost of reaching the right consumer with the right ad message is on the verge of being prohibitive. In embracing Listenomics, is this any time to forget that you have a target audience?

Rick Warren knew his target audience. He is the pastor of the Saddleback Church, a 22,000-member mega-congregation in Lake Forest, California. In 2002, he finished a book. It was well-written, cleverly-packaged — framing our spiritual lives into five simple principles — and inspiring. Big deal. Pastors stitch together old sermons and write books all the time. There are thousands upon thousands of titles on religious themes languishing in warehouses and, more likely, long since remaindered and burned. Some of them, no doubt, were also well-written, cleverly packaged and inspiring. But the Rev. Warren knew what he was about; he had a built-in audience, not only of his own congregation, but of thousands of fellow evangelical clergymen who for a decade had been using Warren's website — pastor.com — for source material in their own ministries.

His email list of 80,000 names initially yielded 1,200 colleagues who were very intrigued to see his trademark concept of "The Purpose Driven Church" extended to spiritual self-help: *The Purpose Driven Life*. So he of course asked them to plug the book from their pulpits . . . right? No, says Greg Stielstra, author of PyroMarketing, who at the time was senior marketing director at Warren's publisher, Zondervan. (And who himself is publisher of the book you are holding in your hand.)

"Instead," Greg recalls, "what Rick said was. 'What do these guys need?'"

The answer, he already knew, was: source material. "So he created a '40 Days of Purpose' campaign," in which all 1,200 pastors received six weeks of sermons to customize or read verbatim, and workbooks for discussion groups within each congregation covering a book chapter a day for 40 days. Oh, and he invoked a clause in his contract that permitted him to buy copies of the $20 book at pennies above cost: $7. These he made available to all the participating congregants with no markup. All 400,000 of them.

That, my friends, is a lot of cheese cubes.

When Zondervan realized it had to print nearly a half million books at near zero profit, Greg says, "The blood drained from their faces." They've since pinked up just fine, however. Those congregants read what they were given and liked what they read. They also bought and recommended *The Purpose Driven Life* to friends and relatives, who have gone on to buy and recommend copies to friends and relatives. Thirty million times, in all. That's the same number of copies sold as *Valley of the Dolls*, *Gone with the Wind* or Anne Frank's *The Diary of a Young Girl*. It is three times the sales of, yes, *Who Moved My Cheese?*

Another splendid example — albeit with a very different target — is the viral video for the pop-music group OK Go. The band members are on treadmills in a choreographed dance number, to a song called "Here It Goes Again" from their album *Oh No*. As this is written, everyone on earth has seen the video at least four times, except for certain remote areas in the mountains of Papua New Guinea, where several tribesmen have seen it only twice. No wonder it took off; it's such a simple, clever idea flawlessly and wittily brought off. And it was the band's *second* mega-viral. In fact, OK Go's video bona fides had been established a year earlier, in 2005, after they decided to record a goofy-but-elaborate boy-band-dance send-up that had amused the audience at a rural England rock festival. The dance cemented their neo New Wave aesthetic, and looked even (stupidly) cooler in a video shot in front man Damian Kulash's backyard to complement their song "A Million Ways."

As my old colleague Kevin Maney recounted in *USA Today*, this was pre-YouTube, but iFilms and other video aggregators were out there, as well as countless pages of the then-new social networking phenomenon, MySpace. Meanwhile, on tour, the band's spotters would look out for the oddest oddballs and nerdiest nerds in the crowd and hand them unmarked DVDs of the video. Naturally, because their lives were lived largely online, these geekazoids rushed home to post the backyard dance wherever they could. That is seeding, and sure enough the germ took root. The backyard dance grew online like kudzu. And CD sales nearly tripled.

So naturally all involved decided it would be a splendid idea to repeat the magic with "Here it Goes Again?" Yeah, right. Think Joseph Heller, and Nicola Tesla, and Van Cliburn, and Michael Cimino (*The Deer Hunter* followed by *Heaven's Gate*), spending whole lives trying to recapture the genius informing their youthful, seminal work. This is the stuff of pathos. Even after contriving to up the choreography ante by doing their dance moves on moving surfaces, and even after having made names for themselves the first time around, the band members faced a daunting challenge. Everybody knew the new routine was even more weirdly captivating than the first one, but, face it: how often does lightning strike twice in the same place? How to make "Here it Goes Again" go again?

Here's how: OK Go appeared on every TV show it could book with the new dance, beginning with VH1 — not as an end in itself, but to generate online interest. There is no rule, after all, that says online success must be entirely self-contained. Though the old media are doomed, that doesn't mean they are irrelevant. On the contrary, there is yet no online equivalent to "TV famous," and with all due respect to Judson Laippley, the "Evolution of Dance" guy, there won't be for a while. Sure enough, the treadmill dance was exactly as novel and fun as all involved believed, and TV viewers began uploading it on a new video hosting site called YouTube. Three weeks later, it had been viewed 4 million times — prompting more TV appearances, yielding more online interest and so on. The band's CD, by then a backlist title becalmed by the recording-industry Great Depression, took off again and now has sold more than 257,000 copies.

Some viral videos actually begin their lives as traditional media and then get dispersed online, like dandelion puffs in the wind. This was the case with the 2008 Cannes Advertising Festival Grand Prix winner for Cadbury chocolates. The soulful drum solo to Phil Collins' "In the Air Tonight," captured the imagination of millions around the U.K. on TV, and then online around the world, not so much owing to the enduring appeal of Collins' music, but mainly to the fact that the drummer was a rheumy-eyed gorilla. Not necessarily wet-your-pants funny, but definitely you've-got-to-see-this funny. Fifteen million views worth.

That's real views, by the way. Attentive views. As my friend Jess Greenwood of London's *Contagious* magazine likes to point out, that 15 million is a whole lot more valuable than a TV audience of the same size. "One," she says, "because they opted in — which is something you can never ever say about a TV commercial. Number two, it came from a trusted friend. They came to it positive, open and receptive. And, number three, when an ad is on TV, the audience may be rattling around the flat putting on a pot of tea."

But do not take that as an unequivocal endorsement. While the ad world may wish to think navigating from the old model to the new one is as simple as making TV commercials and uploading them as virals, Jess Greenwood offers little comfort. For in addition to being just too adorably English for words (pot of tea!), she is also a realistic woman. Just as she gets you imagining the awesome power of a runaway Cadbury's lorry, she deflates a tyre on you. Because seeding or no seeding, the number of ads with multimillion online viewing is very, very small. There was "Gorilla." The treadmill dance. Ronaldinho bouncing a soccer ball off the crossbar for Nike. One for designer Mark Ecco. One for Levis. One for Axe body spray. One for Trojan condoms. One for Smirnoff. One for VW. An amazing (also fake) catch by a ball girl for Gatorade. And not much else. "In terms of branded viral phenomenon," by her reckoning, "I'd say you get four big ones a year."

Sure, for the non Big Four, whatever audience they got was gravy, because the sponsors spent nothing on a media buy. Nonetheless, failing to connect with an audience squanders a lot of time and resources — in exactly the way mass marketing does and digital marketing, almost by definition, should not. God bless Cadbury, but trying to create the next "Gorilla" is a low-percentage bet. Rigorous targeting increases the odds, but not all that much. Which is why Malcolm Gladwell is rich.

He is the author of the bestseller *The Tipping Point*, an exercise in social-networking theory that purports to describe how a handful of extravagantly connected individuals can alter perception and behavior of a grand scale. To read Gladwell's book, with its anecdote about how a few hipsters somehow propelled a déclassé shoe brand like Hushpuppies into a new life as fashion item, is to imagine a rarified cohort of demigods with preternatural sway over vast, concentric social circles .

And who wouldn't want to find these latter-day Saints Peter and Paul? Rick Warren targeted pastors looking to fill up six weeks of pulpit time, and OK Go targeted losers. Narrow as those targets were, they still required a lot of tedious sampling, much of it no doubt utterly wasted. But Gladwell's vaunted influencers seemingly bypass the laborious job of identifying and reaching a broad cross-section of targets to miraculously accelerate the social-network ecosystem. Discover them and you are not sowing seeds; you're planting magic beans, climbing the stalk to the heavens and absconding with the goose that lays the golden egg.

Which, if memory serves, is a fairy tale.

The literature on disproportionate influence, much of it concerning public opinion, styles and fads, predates *The Tipping Point* by about 50 years. I myself first encountered the concept in 1985, while reporting on a room full of chumps who had been lured into a multilevel marketing scam to sell "the credit card of the future" — one which a) you didn't need to make payments on, and b) soon would earn you $9 million per month. One sad sack named Earl, explaining why he was investing hundreds of dollars in a transparent pyramid scheme, told me he himself wouldn't have to do any selling. He'd build an organization of energetic salesmen — he called them "hotdogs" — at the apex of his pyramid and sit back as the base grew and grew. "I'm in it to make money," he told me. "That's my second plan. My first plan is to serve God." Yes, Earl had faith — faith in the Almighty, and faith in the idea that it takes only a few strategically placed hotdogs to fuel exponential growth. And, except possibly for the criminal fraud part, the plan made some sense.

"What we have here is a kind of folk theorem," says Columbia University sociology Professor Duncan Watts, "the kind of thing that everybody thinks is true, but nobody has ever proven — that there is a minority of special people that do a majority of the actual work.... The Gladwellian claim is not only that they have more influence, but that they have vastly more influence."

In 2007, Watts and University of Vermont math Professor Peter Dodds decided to put the folk theorem to the test They constructed a mathematical model (many of them, actually) comparing the reach of especially connected individuals with that of ordinary Joes. The idea was to examine whether those with large, attentive inner circles necessarily radiated more influence to outer concentric circles. Subjects were credited not only for those they influenced, but those influenced by the influenced, those influenced by those influenced by the influenced, and so on. ("Like Amway," Watts says.)

"We made a very simple comparison. If you targeted them with your seeding strategy, how much better would you do versus if you targeted average people?"

The results, published in the December 2007 *Journal of Consumer Research*, were very deflating for those in the business of seeking, or selling, magic-influencer beans. By investing resources into locating the most influential of influencers in a given sphere of influence, Watts and Dodds discovered, "You do a little bit better, but that marginal return is diminishing. Someone who is twice as influential at the first remove is less than twice as influential ultimately." And sometimes no more than an average influencer in the end, and sometimes less than average.

"If you were consistently right (with your targeting) on average you would do better," but not always, and the benefit of a marginal improvement must be weighed against the cost of the targeting process. "If those people cost you twice as much," Watts says, "and they only give you a 10% benefit, it's not worth it."

Yes, Gladwell's backwards reconstruction of the Hushpuppies phenomenon is fascinating, Watts says, but it also was based on a classic logical fallacy: "We notice some things and not others. We don't notice all the things that don't happen. We consistently neglect the non-event. We don't notice all the times when hipsters wear things that don't become popular." There is also the difference between correlation and causation. Yes, some hipsters wore Hushpuppies, and Hushpuppies took off. But many other things happened at the same time, none of which variables were explored by Gladwell. Why assign more weight to the hipsters than to any of the innumerable other factors in play? If one or more of them were actually at the center of the phenomenon, the obsession with human "influencers" is the ultimate red herring. Kind of like the ancients — and poor Earl — attributing all wonders to the gods.

"Stories don't admit to multiple causes," says Watts, who is on leave from Columbia to serve as director of Yahoo's Social Dynamics Group.

"They're simple. They're deterministic. They have simple causes. Because that sounds good to us, we think that's how the world actually works." And because they are stories — per item #6 below — narratives like *The Tipping Point* quickly spread, not thanks to a few specially endowed influentials, but to millions of book-reading ordinary Joes.

So, yeah, it's foolish to bet that influencers within or outside of your natural target audience can shortcut the ecology of WOM transmission. But — if you'll permit me yet another other hand — it's also foolish to forget that they are a part of that ecology. For instance, meet Dan Zarrella, a Boston viral-marketing consultant who would seem to throw in with the "cast a wide net" philosophy: "I think there is a lot to be said about putting your content in places where people can find it," he says. No kidding. I Googled "viral seeding" and found him near the top of the (unpaid) results, the first American consultant listed. Yet he is a one-man shop, who does that work only after he's finished his day job. How did get his name where I could find it?

Google found his web site "popular" because it was heavily linked to. It was heavily linked to, because it was there he posted an article about the science and history of rumors. "How rumors spread," it was titled. The article generated the traffic because it was featured on the main page of Digg, a site which ranks the popularity of content submitted by its registered members. Submitting your own content is frowned upon, especially if you have commercial motives (such as advertising your consultant services) but Zarrella didn't put his own piece up on Digg for a vote. As a Digger for two years, he has become well acquainted with many of its "power" users.

"I've taken pains to make sure there are Diggers with power accounts, and who I have relationships with," he says, and when he posted his article, he used his Twitter feed to notify these key influencers. Sure enough, someone bit, the piece was linked to on Digg and — like a hot rumor itself — got lots of attention. Hence the links. Hence the high Google ranking.

That is search-engine optimization for you, and now you're reading about him here. All of which is to say, while WOM can yield genuine phenomena, a phenomenon isn't necessarily an accident. Although — on what by now must be my seventh or eighth other hand — let's hear it for accidents.

In 2006, the Toronto office Ogilvy & Mather created a remarkable video for Dove beauty products. It was titled "Evolution," the time-lapse story of a plain Jane being tricked out — via make-up, lighting and Photoshop — into a supermodel vixen. Ogilvy put it up on YouTube and sent an email blast about it to its list of subscribers to the Dove Campaign for Real Beauty, urging them to go to the website and check it out. But then, says Janet Kestin, the agency's chief creative officer, "In a stroke of the cosmic, 'Evolution' launched during Madrid Fashion Week, where they decided to ban underweight models, refueling what became known as the Skinny Model Debate. It was perfect: heightened interest in the subject of women, beauty and self-esteem just as 'Evolution' caught people's imaginations."

The incredible convergence offered Dove's PR agency an ideal hook for editors and producers, who wasted not a moment running the video on TV and on their websites. This vastly accelerated traffic to the Dove website and YouTube. About 20 million views later, "Evolution" won the Grand Prix for video advertising at Cannes.

Have a Story to Tell

Dove was, indeed, gigantically lucky that its video landed at precisely the moment when female body-image issues were propelled to the fore. That isn't the end of the story, however, because it neglects the story itself. For three years, Unilever and Ogilvy had waged the Dove Campaign for Real Beauty, with a website, ads (online and off), Dove Self-Esteem Fund and an ongoing education campaign aimed mainly at girls, to inculcate them with a sense of confidence and worth. They weren't lectured that beauty is only skin deep, and that what really counts is our inner selves; children aren't stupid, and they know that how we all look matters in ways large and small. But they were told, and presented with many lovely examples, of physical beauty that doesn't conform to the freakish standards of Hollywood and the fashion industry. From the beginning, it was a fascinating exercise.

From one perspective, all involved were vulnerable to massive eye-rolling on basic hypocrisy grounds; Unilever also makes Slim-Fast, which encourages yo-yo dieting. And it sells Axe and Lynx, body sprays advertised to young males as surefire means to get in the pants of steamin' hot babes who, of course, look like human Barbie Dolls. As for Ogilvy, in a bit of horrifying/delicious irony, it is the ad agency for actual Barbie Dolls.

Ouch. That one stings. Still, there was a second, equally provocative perspective: the subtext of redemption. Even as a leading practitioner in the finding-fault-with-others industry, as a critic I harbored few doubts about the sincerity of all concerned. On the contrary, I believe that the creators of the campaign, at least on the agency side, relished the opportunity not only (for once in their careers) to promulgate a positive, genuinely humane message but also to expiate past sins. Whatever difference that makes. For the purposes of understanding the evolution of "Evolution," it hardly matters whether the ongoing conversation was entirely enthusiastic or entirely skeptical. The only salient point was that it was ongoing. In addition to the cosmic convergence and a jaw-dropping video, Dove had already captured the public's attention and had already developed as intriguing narrative. And what more powerful influence is there on WOM than narrative?

Gossip, for instance. You scarcely have to seed that, because it spreads naturally, very rapidly, and takes root like crabgrass. And why wouldn't it? It's got people, it's got action, it's got drama. They laughed; they cried. As I write these words, I just got off the phone with my daughter, who called specifically to rave about a pan-Asian restaurant in New York where she supposedly had the best meal of her life. What she got back from me — even though I'm sitting here writing about word of mouth endorsements — was basically, "That's nice, dear." This was very different from our conversation only a day earlier, when she called with a semi-scandalous story about two people of our acquaintance waking up in ambiguous proximity to one another, people who had no business being asleep together anywhere. That call, I promise you, had my undivided attention. She had a story to tell — an amusing/appalling one — and believe you me it has been passed along.

What's in it for Us? Not You. *Us.*

Ever been stuck in a conversation with someone not so much dull but militantly boring? The merely dull interlocutor might offer, "They're saying it might rain." The deadly one, though, tells you about his nephew's trombone recital in remarkable detail. He gets especially wound up at the part where he and his wife had to drive around and around for a parking place and then had to walk five blocks and Cecelia was in heels, which he had advised against because of her bunion, but Cecelia was determined to look presentable because her pet peeve is women in public places looking slovenly. "It wasn't that long ago that they wore gloves to church, after all. Honey, didn't your Mom wear gloves to First Presbyterian? I like to tease her because Mother Taylor is a Methodist; she never set foot in First Presbyterian, unless it was for a funeral. Honey, Jimbo McDonald's mom. That funeral was at First Presbyterian, wasn't it? Because ... no, never mind. It wasn't Mrs. McDonald. Remember, Honey, it was Mike Bamford's mom, because she was the in the choir and the whole choir sang. My goodness, but how that choked us up. So to cut a long story short.... "

And you stand there nodding, with a half smile frozen on your face, praying for rescue. That kind of story isn't viral; it's dead on contact.

Why does anyone imagine that just because their family arcana — or brand message — is interesting to themselves, the rest of the world is even remotely concerned? Yet YouTube is bulging at the seams with supposedly "viral" videos that are doomed to containment because they contain nothing of relevance to anyone out there in download land.

Obviously, marketers and content creators and opinionators and politicians wishing to exploit word of mouth do so because they have some interest in being the topic of conversation. It should be equally obvious, though, that nobody else under the sun is interested in, much less motivated by, the desire of others to get noticed. We, as a group, are interested in us. What we value. What we find cool. What we find interesting. What we find surprising. What we find funny. What we find titillating. What we find poignant. What we find disturbing. What we find relevant. What we find useful. You have a product, or a policy, or an opinion you're flogging? What the hell do we care? Babycakes, your problems are of zero concern to us. We're looking out for Number One.

Look at this picture of a character from *The Simpsons*:

Now compare it to the picture, on the back cover, of me. Pretty good likeness, eh? I'd like to boast that I've become such a fixture in popular culture that I've been immortalized/ridiculed on the best TV show in history, but in fact I just used the "Simpsonize Me" engine on the website of the Burger King/*The Simpsons Movie* website. All I did was upload a photo and let the software do to me what Matt Groening might have. Then, needless to say, I sent it to everyone I've ever known. There is, after all, no more rudimentary expression of self-interest than "Hey! Look at me!"

CareerBuilders.com tapped the same instinct with its witty and wonderful "Monk-E-Mail." This was an online tie-in with its TV ad campaign, in which dissatisfied employees were depicted with an office full of chimpanzees as colleagues. Monk-E-Mail allows the sender to record a brief email message, which, when opened by the recipient, is spoken onscreen by a chimp dressed up to look approximately like the sender. About 130 million people found it in impossible to receive a Monk-E-Vision message and not immediately record their own chimp-o-grams to pass along. And OfficeMax's "Elf Yourself" website invited users to upload photos of themselves in order to be turned into a dancing Santa's elf for an emailed Christmas greeting. The cute little opportunity was taken advantage of . . . 193 million times, for the key word in "Elf Yourself" was not "elf."

Yet it is this simple power of self-interest, Jess Greenwood says, that eludes so many viral-video wannabes. Just because they wannabe noticed, no matter how hard you try, doesn't mean we wannabe bothered. On the other hand, she says, "Do something nice for people and they will come to you."

Therein the beauty of widgets, mini-applications residing on personal websites and social-network home pages, that perform some function or another for the user. Weather data, to-do list, analog clock, currency converter, Twitterlex for tracking your incoming tweets, Widsense for tracking your Google AdSense click traffic and revenue. (Much more on this in the next chapter.).

"Viral itself is a bit of an outmoded term," Jess says, "because it was used initially to describe the way content was emailed from person to person. It was before blogs, before social networking, before Web. 2.0. A viral phenomenon now is anything."

Or at least anything useful, relevant and, occasionally, entertaining. In a Web 2.0 world, where zillions of faces inhabit the spaces of their friends and near friends, there is a new ecology. If we're discussing how the flowers of enthusiasm get pollinated (and apologies here again for yet a new metaphor) you don't necessarily need birds and bees flying hither and yon. The flowers come to the bees. Visitors to my website or Facebook page see my CokeTag or my UPS package tracker and take the widget with them. Digg tells them what's popular, and off they go to find it. But never mind the traffic flow. Sites like Digg are the apotheosis of understanding what the target audience values, and some of those things are almost absurdly basic. For instance, celebrity news. If I were, say, Abercrombie, I'd have a portion of my retail site devoted to Britney Spears' blood-alcohol level, Kim Kardashian's booty and other important issues of the day, and a widget on every MySpace page with a headline ticker. Clicking for the full story would take me to the site, where I could also buy the cutest T-shirt ever.

And speaking of "cutest," in their random information gathering, human beings are drawn to superlatives, no matter how dubious. We are all moths, and Top 10 lists are extremely attractive flames. Recall how Dan Zarrella parlayed an article about the dynamics of rumor-mongering into a Digg placement, thence into a high Google ranking. Well, search-engine optimizers such as Zarrella know that some things do better on Digg than others, and lists are at the top of that list. If he had a client, for example, who was a rug merchant, he would recommend that the client compose a Top 10 list apt to interest the tech geeks who are so disproportionately represented on Digg. Something like, "Top 10 Ways Rugs Can Ruin Your Computer. Number 10: Static electricity....."

"Lists in general work very well," Zarrella says. "Beyond that there's the 'Biggest Whatever' or the 'Craziest Whatever' or the 'Geekiest Whatever,'" because our species tends to flock to the first, the biggest, the most. He hastens to add that you must always resist the temptation, having gotten the audience's attention, to sell, sell, sell. Naked commercialism does not sit well with Diggers, or most anyone else online. But that doesn't mean they won't listen to a rug expert talk about rugs, even if there is an implicit commercial benefit to the interaction.

"If the content is good enough," Zarrella says, "people are willing to look past that."

Exactly. So please ignore the fact that the Top 10 list you are now reading is a cheap gambit for me to sell books. Love it for its deca-delicious self.

Behave Yourself

As you'll see a couple of chapters hence, I recently spent an entire year happily making life miserable for a certain gigantic cable juggernaut. (Hint: it rhymes with "Comcast.") A few weeks into my jihad, a guy writes to ask me if I'm on the take from some other telecom: "Bob — can you gives us a good faith disclosure that you are not receiving compensation or inducements from another provider of video & broadband services? — Oliver Boulind, New York, NY"

These are fighting words, and I'd punch this guy in the snoot for impugning my integrity — except it's not an entirely stupid question. There are a lot of pimps and whores out there, and some of them have walked the streets of the internet for Comcast. Here's a little tidbit from a blogger called "Lutchi," who at the time was selling herself for $5 a post:

Are you still looking for the best internet service provider ? then, don't look too far because COMCAST is here and ready to serve you every-thing that you need.Comcast, is now the leading provider of high speed internet access, and Comcast Digital Voice. Their broadband internet connection is super fast and even beat that Verizon DSL in numerous speed tests. And if your tired dealing on paying big on your recent phone plan then you might need to consider using their digital Phone service. Right now, you have 3 plans to choose from starting from $39.95 to $54.95 a month. You can get unlimited local and long-distance calling, three way calling/ Caller ID/waiting, voice mail online, and it works *with your existing phone. You can enjoy all their cool features without sacrificing anything because you still get what you got from your previous phone services.*

I have to tell Joe, about this phone service because it is a lot cheaper and I am hoping they are available on our area. So guys, why not visit them today and check if Comcast Digital Voice is available in your area. And if you have any questions about anything regarding their services please feel free to contact the customer support for more information.

Comcast says this blatant example of pay-for-post was not a headquarters undertaking, but the work of an unscrupulous subcontractor. Maybe so. What it certainly was was sleazy. Also pathetic and so unnecessary. If a service provider wants good buzz, it can get it in exponentially greater volume for free. All it has to do is provide good service. But good buzz is not a commodity. Word-of-mouthster Andy Sernovitz, ordinarily the world's most affable guy, gets extremely irritable when he encounters people trying to buy word of mouth as if it were column inches or airtime. He thinks this has something to do with marketer desperation, the realization that their days of dictating messages are over, leaving them feeling lost, impotent and vulnerable to the entreaties of vendors claiming they can influence the conversation online.

"There's giant pressure on companies to do something with Web 2.0," Sernovitz says. "So they hire someone to give them a buzz campaign, a Facebook page or a 'viral' video. Because those are easy to buy. But true customer engagement and real conversations are hard to do. They're not about money, and your agency can't sell you one."

But, my, how they try. PayPerPost.com, for instance, a subsidiary of Izea Social Media Marketing, does plenty of business promising advertisers "a vehicle to promote your Web site, product, service or company through the PayPerPost network of over 50,000 independent bloggers." Meantime, it recruits the bloggers thusly: "Advertisers are willing to pay for your opinion on various topics. Search through a list of Opportunities, make a blog posting, get your content approved and get paid. It's that simple."

Yep. It is simple. So is a $20 hand job.

If that analogy doesn't crystallize the issue for you, perhaps it will be helpful to consider the largest payer-per-post in the world. I refer, naturally, to the Communist Party of China, where hundreds of thousands of e-moles are deployed to smack down party or government critics online at 50-cents per ringing endorsement. The critics, meanwhile, go to jail. So if you're tempted to bribe stooges to fight your PR battles, do consider the company you keep.

First cousins to PayPerPost are outfits that recruit ordinary folks to shill for clients not only online, but everywhere: bars, church, airplanes, hair salons, book groups, school cafeterias, etc. These people don't get paid cash money; their compensation is the thrill of having their opinions count, plus free samples of whatever crap they happen to be flogging to the poor saps in the room with them — saps who mistakenly believe they are having a conversation, only to discover (or not discover) that they're subjects of a sales pitch. "Sending out secret agents to annoy people," as Sernovitz puts it.

For obvious reasons, under the Word of Mouth Marketing Association code of ethics, stealth is strictly forbidden. When it's unsuccessful, it injures the reputation of the client; when it is successful, it is quintessential deceptive marketing. In the United Kingdom, it is a crime. But how often deception occurs, and who is responsible, gets a little squishy. Consider the most prominent vendor in this business, BzzAgent, which so far has no legal woes in the U.K., or the U.S..

"Deception occurs when it is a secret, paid endorsement," says founder and CEO Dave Balter, who at any given time has 80,000 to 150,000 agents buzzing about Listerine Whitening Strips, Grey Poupon, Pledge and a bunch of other products you've probably never brought up in conversation. But Balter is quick to point out that 1) his "agents" are paid no money (they get only product samples, reward points and the pride of membership), 2) they are obliged by company policy to disclose their affiliation and, 3) they are free to pan a product as well as endorse it. The company diligently trains and constantly reminds its people not to operate in stealth. "We have to untrain people who thinks being secretive is really cool," Balter says. "If you do not disclose, we kick you out of the system."

Of course, Balter also readily acknowledges he has no way to tell whether any of his agents are doing so. All he has to go on is their own contact reports, and the firm belief that "Being an agent is a badge of honor." Well, maybe it is; humanity is sometimes a sad affair. But if it is human nature to want to be part of something and to have your opinion count, isn't it equally human nature not to want to listen to unsolicited commercial endorsements from relatives, friends and total strangers? And isn't it therefore more humanly natural not to want the lady in the next salon chair, or the guy at the bar, or Aunt Clara — when you spontaneously talk up Brand Whatever — to tell you to piss off?

I certainly would. I mean, doesn't everybody just love learning that a conversation is really a commercial, just as everybody loves being invited over to the Crenshaws for dinner, and then, during the dessert course, watching them set up an easel and start drawing Amway circles on it. That makes you feel *soooooo* special. But, more to the point, if the power of word of mouth resides in the trust we all put in our friends, neighbors, relatives and colleagues compared to commercial messages of any kind, disclosure immediately unplugs it. A bzz-kill, you might say. So while I'm inclined to believe what Dave Balter tells me, I don't believe a word of what his BzzAgents are swearing to him. Still, because of its strict official guidelines, BzzAgent remains a member in good standing at the Word of Mouth Marketing Association — at least until some wayward agent forgets to disclose his free case of Mahatma Gourmet Rice to the guy next to him on the Boston-Washington flight, the lawyer for the FTC. Meanwhile, another big player called Tremor is ineligible for WOMMA membership, because it does not require its network of teenage shills to tell their fellow teenagers a thing. Tremor is owned by Procter & Gamble, which should be ashamed of itself.

Disclosure is a classic damned if you do/damned if you don't proposition. But especially if you don't, because most likely you will be caught, and getting caught sucks. These episodes crop up periodically and never fail to damage the reputations of the perps. In 2002, Sony Ericsson hired 60 actors, equipped them with T68i mobile phones and deployed them to various cities in the guise of tourists asking other tourists to snap their pictures with camera phones. The *Wall Street Journal* duly reported this exercise in guerilla marketing, and very quickly Sony Ericsson had a backlash on its hands.

In 2006, a couple named Laura and Jim crossed country in an RV, bedding down each night in a Wal-Mart parking lot, communing with employees and shoppers and blogging about the experience. Soon they were unmasked as creations of Walmart's PR agency, which had also created two other "flogs" — fake blogs — for its client. (One of them was attached to Working Families for Wal-Mart, a fake grassroots organization also created by the agency to deflect criticism of the world's largest retailer.) In the same year, a blog called The Zero Movement was formed, ostensibly by a random lazy guy, to embrace the ethos of new Coke Zero — i.e., in praise of doing nothing. Ha ha. It was quickly revealed to be a Coca-Cola Co. undertaking, and Coke Zero took a shellacking online (and at retail.). A few months later, Sony pulled the same stunt in support of its PSP portable videogame device, using a viral-marketing agency to fake a blog by a supposed hip-hop artist named Charlie.

It was a fraud, but it sure made its mark in the blogosphere: "Transparent, insulting, idiotic, and ineffective," wrote one blogger, who I'm quoting because he was one of the more civil ones. Also unamused by such behavior is the Federal Trade Commission, which late in 2006 announced that marketers whose agents fail to disclose their affiliation will be subject to fine.

Remember Siegfried and Roy. Especially Roy.

Just when you think you do have the tiger by the tail, do not ever forget, the tiger can turn against you. Word of mouth is a wild and dangerous beast.

I've already mentioned *Ishtar*, the Dustin Hoffman/Warren Beatty vehicle that crashed and burned in 1987. Think, too, of the Salem witch trials through which mean-spirited gossip mutated into community-sanctioned murder. Or the bank runs of the 1930s, when panicked depositors heard rumors about losing access to their savings and by queuing up at teller cages created a self-fulfilling prophecy. Or the McCarthy witch hunts of the '50s, built on nothing but whispered denunciations about supposed disloyalty. Or the current conspiracy theory, considered fact among more than half the population of the Arab and Muslim worlds (according to the 2006 Pew Global Attitudes Project), that the attacks of 9/11 were not an Al Qaeda plot but a U.S. or Israeli one.

Or, just for the diabolical fun of it, try Googling "Procter & Gamble" and "Satanism." Decades of bizarre rumors, rooted in P&G's vaguely creepy moon-and-stars logo, have connected the maker of Tide, Pampers, Crest and Oil of Olay to the Church of Satan. This culminated in the urban legend that the company president appeared on the Sally Jessy Raphael/Merv Griffin/Phil Donahue program in 1999 to declare corporate sympathy for the devil. You can even look up the date. It was March 1. And when Sally/Merv/Phil asked him if the Satan connection would hurt business, the executive replied: "There are not enough Christians in the United States to make a difference."

Obviously, none of that really ever happened. Procter & Gamble is not a tool of Satan. (Duh. That would be Microsoft.) But many cling stubbornly to the belief, because they heard it from their hairdresser, or their cousin the Amway distributor, or their pastor.

The oft-cited power of WOM, of course, is that we all tend to deem more credible information passed on by people we know personally. Unfortunately, this is true of misinformation, as well, and misinformation is extremely difficult to eradicate. P&G fought back against rumor mongers, in some cases by successfully suing them, and using the media to set the record straight.

Which may have made the problem worse. A 2007 study by University of Michigan social psychologist Norbert Schwarz demonstrated that attempts to correct misinformation not only tend to reinforce people's false beliefs, but to attribute the baloney to the very authoritative source trying to clear things up. The particular case researchers focused on was a Centers for Disease Control flier that aimed to debunk myths about flu vaccines. The Michigan study revealed, however, that the very myth-busting process had the opposite effect. People came away believing that they'd been warned by the CDC to avoid flu vaccines.

That cognitive quirk is no doubt very reassuring if you're trying to tell the world, say, that Saddam Hussein had weapons of mass destruction and was behind 9/11, but if you're not yourself in the Big Lie business, the persistence of myth is reason for pause. You do not want word of mouth working against you.

In April 2009, two knuckleheads then employed at a North Carolina Domino's made a video, in which one of them stuck mozzarella in his nose before placing it on a lunch item he was preparing, then farted on a salami slice headed for the same delicious sandwich. Within three days (and just under a million YouTube views of the video prank), the chain was forced to put up its own YouTube video denouncing the stunt and certifying that salami-slice farting is not a company-sanctioned sandwich-making technique. Ten days after *that*, the response had amassed a grand total of 600 views. Youch. Or consider the plight of poor, poor Microsoft, which, as previously noted, launched its Vista operating system in 2007 only after previewing it with, and building a community of, some 5 million beta testers.

"A new release of Windows is always a highly anticipated product," says David Webster, general manager for brand and marketing strategy. But for the first time, Vista was released into an environment of Gizmodo, Slashdot, Engadget and the other blog feeders of digital wildfire. Word of mouth activity this time around, Webster says, seemed "to be an order of magnitude larger."

At first, as word-of-mouth maven Sernovitz correctly observed, the online attention served Microsoft rather well, and the new product flew off the shelves. Twenty million copies of Vista were sold in the first month, double the pace of the XP launch five years earlier. Alas, it was also a double-edged sword, as large numbers of those early customers ran into big problems: hardware incompatibilities that froze printers and other peripherals, onerous Digital Rights Management protections interfering with CD burning, memory demands slowing their systems to a crawl, too many prompts interrupting basic functions and so on. Quickly, the mood of the crowd started to turn, and a part of the rah-rah community devolved into something of an angry mob. Not a small one, either. If you Google "Vista nightmare," you get more than 2.4 million hits along the lines of this: "For all of you out there who like me bought a new computer with VISTA as the opporating [sic] system. GOD HELP US ALL!"

Internal Microsoft documents, surrendered in the discovery process of a class-action suit, even revealed Vistangina at the highest levels of the company. Mike Nash, vice president for Windows product management, complained that he bought a supposedly "Vista-capable" Sony laptop only to discover that the machine's video chip was rendered inoperable. "I now have a $2100 email machine," he said.

So mark down Mike Nash as "disappointed." Also Adolf Hitler. In a hilarious YouTube clip featuring fake subtitles from a scene in *Downfall*, der Fuhrer is seen in a tirade against his general staff for ruining his computer with Vista: "You swindled the wrong guy!" he shouts, at least in the subtitles of the German dialogue. "Assholes!... XP always served me good. I really should have stuck with XP, like Stalin!"

More than 200,000 people have viewed that video, (http://www.you-tube.com/watch?v=bNmcf4Y3lGM) which is laugh-out-loud funny only if you accept as a given that Vista totally sucks — a sentiment that quickly dominated not only the blogosphere, but the so-called mainstream media, as well. *PCWorld Magazine* declared Vista the "World's Biggest Tech Disappointment of 2007" and InfoWorld rated it the "#2 Tech Flop of All Time." (Right behind the industry's terrifying inability to provide net security.) Thus resounded the echo chamber, to the point that Vista's shortcomings took hold in the conventional wisdom.

The phenomenon is hardly unique to Microsoft. It similarly afflicts a global brand called the United States of America. You and I might see Brand U.S.A. as a bastion of democracy, opportunity, tolerance and individual freedoms the likes of which history had never previously seen. Elsewhere — especially in the Middle East and the rest of the Muslim world — our image is a little less rosy. There we are occupiers. Infidels. Libertines. Zionists. Imperialists. Evildoers. The Great Satan. You know the old saw about, "I don't care what you say about me as long as you spell my name right?"

It isn't true.

Pray for Serenity

> *God grant me the serenity*
> *to accept the things I cannot change;*
> *courage to change the things I can;*
> *and wisdom to know the difference*

So the wild beast turns on you. Then what?

I can tell you what the United States government does. It spends hundreds of millions of "public diplomacy" dollars trying to sweet-talk the beast. Here, kitty kitty. Here's al Hurrah, a propaganda satellite-TV channel! Here kitty, kitty. Here are some mawkish TV commercials featuring smiling young Muslim Americans in headscarves! Here, kitty kitty. Here's a taste of our freedom!

The beast, needless to say, isn't impressed. Because the beast is concerned only with America's policy *vis-à-vis* Israel and the Palestinians, and with our propping up of tyrants like Egypt's Hosni Mubarak, and with our devil's pact with the Saudis. Trying to eradicate animus, while maintaining the underlying policies, is a fool's errand. As the Michigan study concluded, the more we try to disabuse the perception, the more ingrained the perception becomes.

So let's now revisit the Vista problem. If you happen to be Microsoft, and you realize that the conventional wisdom is suppressing sales of your core product, and you won't have a replacement for that core product till at least 2010, there are a number of ways to react. One, per the prayer, is the "serenity" to accept your fate, which could also be called "surrender." Another is "courage," but which could be interpreted as "recklessness." But perhaps there's a middle path, one not specified in the verse (made famous, if not actually written, by Reinhold Niebuhr): bite your lip, try your damnedest to change the conversation and try even your damnedester to not seem ridiculously defensive in so doing.

"I think the new reality of launching high-tech products with a very active online audience is there is a cycle you go through: where you're the greatest thing since sliced bread, then they can't wait to tear you down," Microsoft's David Webster told me. "It's a very, very common arc at this point. For us it's an arc that's playing out against an absurdly large population. No matter what you end up delivering, there's going to be a percentage of that population that is not happy."

That's obviously true. It's also a bit of blame shifting. Webster neglects to add that if the product is terrible, the "unhappy" percentage will grow. While he points to internal data reflecting 89% customer satisfaction with Vista, I tend to multiply the disenchanted 11% against the 140 million units sold. That math yields 15.4 million people who paid 200 bucks only to be stymied, frustrated and plenty pissed off. But let us not quibble, because Webster says the corporate goal is not to tell those 15.4 million people they are wrong, but rather to minimize their influence on the billion or so computer users who haven't bought Vista — the vast majority of whom, he says, will never run into the problems the early adopters did, because early adopters tend to be sophisticated users with rarified demands.

"The enthusiast crowd became so vocal, that it sort of bled into the mainstream audience. And my mom heard that Vista was not a good product," Webster says. "I'm not saying they're wrong in reaching their conclusions. I'm saying their conclusions are not good predictors of what my mother's going to want. The challenge for us is breaking through, changing the frame, changing the conversation a little bit, giving people an excuse for giving Vista a try."

Which, a half-year into Vista's downward trajectory, became precisely the Microsoft corporate strategy:

1) to acknowledge issues with the product,
2) to assert that those issues have been exaggerated by the ill-informed and by the very nature of the product-adoption cycle and
3) to tempt those non-customers tainted by the negative conventional wisdom to try the product for themselves. "To try to get consumers to be a little skeptical. I know they're skeptical about *us*. I just want them to be skeptical of what they've heard from other people."

Hence, in August 2008, "The Mojave Experiment." With the help of Indianapolis ad agency Bradley & Montgomery, Microsoft located 140 San Franciscans who had heard the negative buzz about Vista, but never tried the product, and brought them in for a demonstration of a new operating system — one they were told was called "Mojave." After a 5-minute demo (on a properly equipped mid-range computer), they were asked their impressions. Sure enough, they loved this new Mojave. They rated it an 8.5 on a scale of 1 – 10, compared with their sight-unseen 4.4 rating of Vista Then, with the hidden camera rolling, they were informed that they were actually sampling new Folgers Crystals, with the fresh-brewed taste. I mean Vista.

> *"Are you serious?"*
> *"Son of a gun. You've got me."*
> *"Actually, it's totally different than what I'd heard it would be like."*
> *"I'm getting it!"*

Uh huh. The question is, is Microsoft getting it?

While it has created some patches to deal with some of the compatibility bugs and leaned on computer manufacturers not to bundle it with underpowered machines, the company is deluding the consumer, and itself, to represent the nasty buzz as ravings of a few ornery loudmouths. This is how I characterized the problem in my Ad Review column:

> *As its parting shot on the Mojave Experiment website, we see a subject being informed that his "Mojave" was actually Vista. "But why is it faster?" he asks, and this is supposed to be the Aha! moment. Epee to the heart. "If it doesn't fit, you must acquit." Except that his happens to be precisely the right question. As a Digg commenter named Market-mule explained: "An optimized installation of an OS, by an optimized technician, on optimized hardware, under optimized conditions doing an optimized set of tasks. With conditions like that I could make a rose out of excrement and win first place in a flower show."*

Roger that, Marketmule. You didn't even mention the human dynamic: subjects predisposed to pleasing their gracious hosts.

Webster says the success of The Mojave Experiment will be measured by the response of non-customers. "We just needed to give them a reason to take another look." But it was a low-percentage bet. For one thing, the campaign created an immediate online backlash on the very megablogs that had bedeviled Vista from the start. Implying that your critics are full of shit (as I can attest, alas, on both sides of the equation) is a pretty surefire way to radicalize your critics. Furthermore, woe betide Microsoft if consumers buy Vista on the strength of this campaign and go on to have a bad experience. They will feel not only disenchanted but positively flimflammed, and rush into the camp of the radicals, compounding Vista's problems exponentially. Even for a company that commands 91% of the operating system market, that could be a devastating blow.

A brand with a history of flawed software and predatory business practices does not have much goodwill to squander. "Thuggish" and "incompetent' are bad enough without adding "con artist" to the bill of indictment. Microsoft's middle path between surrender and recklessness may sell some more copies of Vista, but at great risk.

Me, I'd have opted for serenity.

THE WIDGETAL AGE

SOME GUY LIVES IN ALBUQUERQUE, which is great, because it is sunny and really convenient to Vista Encantada and Hoffmantown. But he has relatives in Denver, a limited budget, a lot of outstanding family obligations and a 7-hour, 450-mile gulf between them. Then, one hot and dry Thursday, he's sitting at a computer, and it goes...."DING!"

An icon on his desktop has some breaking news: a special Albuquerque-Denver fare on Southwest Airlines for $49 each way. It sends him that alert because he's asked for it, by downloading the Southwest "Ding!" widget. Most of the time it just sits there, apparently idle, a tiny Southwest logo on a tiny Southwest tail section reminding him, at some extremely low level of consciousness, that Southwest exists. But then it dings, and then he clicks, and, because Uncle Ramon and Aunt RuthEllen simply must be dealt with, then he books.

"After the first year, we hit the 2 million mark for downloads," says Paul Sacco, senior manager for online strategy and development at Southwest Airlines. "And it's still performing." In the third quarter of 2008, Ding! generated 10 million clicks.

Wanna get away ... from The Old Model? Look no farther than the widget, the mini-software applications downloadable to browsers, desktops, social-networking pages, home pages and your mobile phone. It may not be the Holy Grail, but it's arguably pretty damn grail-ish — maybe the highest expression so far of online marketing in the post-advertising age. And though it is very much on the cutting edge of Web 2.0, it is based on a hoariest of principles. In fact, to be properly visionary on this subject, you must begin by looking way back to the future.

For the past half century (and for about five more minutes) TV advertising has been at the apex of marketing communications. Then, in no particular order, newspapers, magazines, radio, out-of-home, direct mail, point-of-purchase, collateral (brochures, for example) and — in the murky, mucky darkness at the very bottom of the deepest abyss of marketing prestige — advertising specialties. Such as a ballpoint pen emblazed with your insurance agent's logo. Or wall calendar, fridge magnet, coffee mug, yardstick, foam beer-can sleeve, ashtray, key fob, emery board, pocket diary — any cheap giveaway item meant to remind the consumer of you every single time she measures fabric or swigs a Pabst or files her nails.

Not that the 30-second spot represents high culture, exactly, but it's hard for mere words to convey how déclassé is the advertising-specialty niche. Still, I'll try: They are the white-belt/white-shoes Full Cleveland of marketing. In a digital world, advertising specialties are as analog as you can possibly get.

Until they go digital. Branded widgets are the refrigerator magnets of the Brave New World. These compact, portable little software apps — from video-players to countdown clocks to coiffure simulators — are inexpensive to distribute, free to the user and (often enough) distinctly useful. At a minimum, they carry an ad message wherever they go.

That's at a minimum. At a maximum, the widget is something like the magical connection between marketers and consumers, not only replacing the one-way messaging long dominated by media advertising, but vastly outperforming it. Because online the link is literal and direct, and along its path data of behavior, preference and intention is left at every step. Oh, and your target consumers actually go out searching for your branded jimcrack. Oh, and they display within easy reach. Oh, and they pass copies along to their friends and associates. Oh, and because they've been turned on by a friend, they are hospitable and receptive recipients. And, oh, in case this didn't quite register the first time I mentioned it, the barriers to entry are preposterously low.

"The money is a joke," says Hillel Cooperman, ex-Microsoft big shot and founder of a small Seattle software-development shop called the Jackson Fish Market. "It's a rounding error in the marketing business."

Ditto that, says Michael Lazerow, CEO of branded application house Buddy Media, New York, especially when it comes to the cost of advertising (or, as he calls it, "app-vertising") the widget itself. "It is so cheap. This is the steal of the century."

That's because 500 million social-network users, each generating 1200 page views per month, represent 600 billion monthly opportunities for an ad impression delivered each time. Hence, almost everything is a remnant and, "You can buy inventory for basically nothing."

Time for a Makeover?

Of course, that invites online marketers to embrace another throwback concept: an endless fusillade of mass messaging with no distinct target — which is pretty much what digital marketing was supposed to be the solution for, wasn't it? But more on the economics of widgetry to follow. For the moment, let's look at some examples that demonstrate why, at least for the time being, it represent the very apotheosis of digital marketing. Remember my pal Jess Greenwood of London's *Contagious* magazine? Here's how she sums up the widget's value "It's like a basic unit of utility. The marketing becomes part of the product." Such as:

- Miles, a 3D desktop avatar who looks like a refugee from the *Teletubbies*, but who resides on your desktop to encourage (i.e., nag) you to run, and keep track of your progress via the astonishing Nike Plus technology. He also keeps you apprised of local weather, running events, promotions. And he organizes your RSS feeds, so you can easily download to your iPod.

- UPS Widget. This guy looks like Miles's tan cousin. He allows you to schedule and track shipments worldwide with a click or two. If you are any sort of frequent shipper, why wouldn't you install him on your desktop?
- CokeTags is a Facebook app that displays your favorite links, allowing you to itemize your online self.... and keep track of who is following the trail of self you blazed.
- Steepandcheap.com is an alert mechanism from the Backcountry.com catalogue advising of special deals — mainly loss leaders — that draw users into the online-shopping experience. It's essentially like the Southwest Ding!, and works because the audience is as much a social network of the outdoorsy as a list of gear customers.
- InStyle.com's The Hollywood Hair Makeover allowed users to lift the coiffures of Jennifer Aniston, Cameron Diaz, et al and superimpose them on their own photos — for fun and/or to show their stylist. A superficiality bulls eye.
- Johnnie Walker. So you're in Singapore, old enough to drink in bars and young enough for that to be a lifestyle. Download the "Jennie" widget, and there is a totally cute avatar who guides you to the coolest saloon events and then, if you're half in the bag, safely home.

Kind of hard to imagine users installing and using these ingenious apps and not being appreciative of the sponsor every single time — appreciation no banner-ad placer could ever hope to achieve. As high-tech entrepreneur and former digital-marketing analyst Peter Kim puts it, "When you can combine utility with the purpose of your brand, that's the opposite of why people hate marketing. Instead of fooling them with the old brand-marketing song and dance, it's not a promise, it's a reality: 'This is what the traffic is like. This is what the weather is. This is what the stock market is right now.'"

Yeah, I'm grateful to John's Hardware when I kill insects dead, dead, dead with my free fly swatter, and I'm grateful to Johnnie Walker when it helps me find the best Singapore bar and find my stumbling way back to my bed, bed, bed. That's the kind of dynamic that gets folks excited, folks such as *Newsweek* and GigaOm's Om Malik, who each declared 2007 "The Year of the Widget." And why? Because the marketer essentially gets to set up shop where you live, work and play.

"Inside the destination and the context people are already engaged in," says Niall Kennedy, founder of consultancy Hat Trick Media and host of the annual Widget Summit in San Francisco. "I liken this to a small Cincinnati retailer setting up shop in all the cities of the world — instead of waiting for people to come to visit Cincinnati."

Yet in 2008 the entire segment amounted to approximately $100 million. That's not nothing, but even in the midst of economic implosion, that's a sum even an endangered species like NBC Universal can shake out of the sofa cushions. Which drives software developer Hillel Cooperman right up the wall. As he ventures out to AdTech and other prominent forums about marketing's future, he is confounded by his inability to capture advertisers' attention. Even as they bemoan the ongoing collapse of traditional media, and worry aloud about where to spend their money, he says, it's as if he's invisible. "And I'm jumping up and down. 'Hellooooo!. Over here!'" he says, the frustration ringing in his voice. "All the stars are aligning. Everybody around me says you're in the right place at the right time. Yet it's still like pulling teeth."

Hmm. Magical connection. Pulling teeth. Those two images are hard to reconcile, but let's do try. There are plenty of reasons marketers have been slow to exploit the possibilities — and why, no matter how grail-ish it can be, the widget's place in even a fully digital marketing economy may have a relatively low upper limit.

"The whole concept of a widget is just misstated or overblown," says Ben Kunz, director of strategic planning for Mediassociates, a media planning firm. "It's not the channel, it's what you do with it that it's important."

Caution! Another List Ahead!

Kunz actually likes widgets a great deal; he's just queasy to hear them oversold given what he says are their inherent limitations — not the least of which is the difference between engaging with a piece of software and engaging with the sponsoring brand. "There's a lot of hyperbole out there about engagement," he says. For instance, if Schick Quattro sponsors a widget that lets a guy embed his face on the hunky body engaged in a pillow fight with two barely legal teens — and it did — does this carry over to razor purchases? "Does throwing pillows at each other," Kunz asks, "really influence anyone?"

Even if the answer is yes, there are plenty of other issues to consider.

1) Non-standardization. There are lots of incompatible platforms: desktop, iGoogle, mobile, Facebook, MySpace, etc. Pending software-code universality, you must create a half dozen or more versions of every widget.

2) Dubious relevance to low-interest categories. What makes perfect sense for Johnnie Walker and Nike may not necessarily apply to Charmin.

3) Cost. While, as Hillel Cooperman correctly observes, the cost of creating a widget is enticingly small, and the cost of distributing one low compared to media advertising, the price tag is also typically open ended. Marketers can be socked with up to a $5 fee every time some Courtney or Madison embeds their widget on her MySpace page. So if you get lucky, you could also get unlucky with a whopping distribution bill. "Open-ended" is hard to budget for.

4) Scale. There is only so much space on a desktop, or a Facebook page, or a mobile phone screen. As Ben Kunz observes, "Sure you can give them utility, but there are only so many slots for that utility. In my world there may be 100 things that I use a computer for. So conceivably you can create a widget for each one of those things, but you've limited the inventory." Which means the vast majority of marketers are shut out the vast majority of the time.

Then there is The Great Widgetry Schism, a fundamental philosophical difference among users and developers as to what the technology is best suited for. In commissioning a widget, do you wish to be all the rage with those notoriously fickle Courtneys, who create viral sensations that spread far and wide but quickly peter out? Or do you shoot for endurance, residing on home pages and desktops perhaps in perpetuity? The bias among widget shops seems to be entertainment over utility, essentially using widgets much like ads: to briefly get users attention and then to start over when that attention wanes.

"It's a campaign model," says Liza Hausman, VP of marketing for Gigya, the largest widget distribution agency. "Advertisers are still going to have to move the needle in a particular time-frame. There are people who are looking at widgets as CRM [customer relations management] or long-term dialogues. That's not where we focus."

Hausman says that this model also conforms well with consumer behavior, especially among the habitués of MySpace, etc. There, she says, the user's page is an ever-changing expression of self, which "self" often is expressed in the form of a showcase for the user's latest discovery. In short, says Hausman, "users like to update their pages." Thirdly, if you assume that utility-based widgets tend to reside on desktops, versus social-networking pages, the utility imperative comes at the expense of virulence.

"That is a 1-to-1 relationship," Hausman says. "That widget is seen only by the person who put it there. Those widgets help you get through the day: news weather, info," compared to social-network-page widgets residing "where people are putting a public face on their world. And the widget there has a 1-to-many exposure."

That argument would seem to be backed up by data. A study by online market-research firm Marketing Evolution found that return-on-investment from widgets increases in approximately direct proportion to virulence. The study of campaigns from Adidas and video-game publisher Electronic Arts within MySpace communities found that 70% of the ROI was attributable to consumer-to-consumer proliferation.

The consulting firm calls this the "momentum effect," and clearly the momentum is a function of the kind of sharing that, say, the Southwest Ding! doesn't much enjoy.

"I have no vested interest," says Marketing Evolution CEO Rex Briggs. "But I do tend to lean that way, mainly because we know there are decay curves to advertising. Keeping it novel and fresh generates a larger response. Part of it is the ability and desire to pass along what you're saying. Something new makes it newsworthy, worthy of passing it along to others, and there's value to that."

On the other hand, companies like Gigya absolutely do have a vested interest and a structural bias against the enduring-utility model: duration militates against repeat business. If a client successfully lands a widget on a zillion desktops or social networking pages, and it stays there, the client has far less incentive to commission subsequent efforts., which, obviously, is bad news for the software designers and distributors. Furthermore, the problem with entertainment widgets such as games is the same one that afflicts any form of viral marketing: Virulence is hard to achieve. No matter how many Courtneys are out there, it's really hard to be the Next Big Thing that, however briefly, captures their imagination. As Cooperman puts it, "Games are just like music and movies and books. It's a hit-driven business. I think it's fair to say nobody knows how to make a hit in any of those industries."

Whichever side of the schism you embrace, widgets offer advantages that hardly any other marketing tool can match — chief among them: portability. Can't get people to visit your website? Once they visit, you can't lure them back? Try the amazing new Website-in-a-Can! It's compact! You can store it on your desktop, your Bebo page ... or you can fold it up and put it in your toolbar! *Funtime, partytime, anytime!*

Just One Little Snag

And it really, really works! Logging data like a website, enabling direct commerce like a website and being a destination like a website, only without the user actually having to leave the virtual house. She merely goes to the cupboard and opens up a can. For free, of course. Because Website-in-a-Can is so cheap to produce, folks can just give it away. And lots of really cool stuff with it, including — to cite my favorite example — entire feature-length movies. In fact, permit me please a brief departure from modern advertising specialties to give you a sense of widgetry's larger implications. If an airfare-alert ding doesn't much impress you, spend a little time with Ted Leonsis and his brainchild SnagFilms.com.

You know Leonsis: an internet entrepreneur before there was even a worldwide web, he merged his company with fledgling AOL and several hundreds of millions of dollars later retired to philanthropy, start-ups — such as leading widget syndicator, Clearspring — and his hockey team, the Washington Capitals. In three different ways, you might say, he is a what's-wrong-with-this-picture sort of tycoon.

First of all, to look at him, with his stocky build and goatee, you'd be forgiven for seeing some sort of central-casting heavy, a Greek Town kingpin right out of *The Wire*. In actual fact, though, he is a soft-spoken visionary, with smiling eyes and charitable heart. Secondly, he has a gift for looking at large systems and homing in on their structural flaws — and then finding solutions. That's why, on the day I see him in May 2009, he's in a very good mood.

The Caps have come from behind to defeat the New York Rangers in the first round of the Stanley Cup Playoffs. His players are wandering around their Ballston, Virginia training facility after practice, adonises in flip-flops. Over the next 10 days, they will extend the mighty Pittsburgh Penguins to seven games, evidence that the frustrating franchise purchased by Leonsis in 1999 had broken its pattern of big ambition-heroic effort-playoff impotence. Thus, like so many other characters and settings mentioned in this volume, the Caps represent something larger. These boys aren't merely highly-paid skaters with funny accents; as a group they represent a microcosm of Leonsis' special gift for problem solving. In the Caps' case, he stopped paying huge dollars to established superstars who didn't quite mesh with the team, and began instead to invest in youth, especially Russian prodigy Alex Ovechkin.

All of which gets to the third aspect of the "what's wrong with this picture" puzzle. A few years back, Leonsis started *producing* pictures — documentary films, to be precise — and he quickly found out what was wrong with them. The film was *Nanking*, documenting the shocking prelude to world war: 200,000 deaths and 20,000 rapes in China's then-capital, inflicted by troops of the Japanese Imperial Army in 1937. For whatever reason, the U.S. distributor thought it would be a good idea to premiere the film in New York during the Christmas holiday. "It's not a happy, family comedy," Leonsis says, still shaking his head a year and a half later and trying to imagine what the distributor imagined the holiday audience's reaction would be to a subject such as *Nanking*: "'Hey, let's go ice skating at Rockefeller Center and let's take in that Chinese-holocaust film!'" Needless to say, not many moviegoers actually reacted that way — despite an ad in *The New York Times* which, Leonsis recounts with still more incredulity, cost more than the distributor would have recouped had it sold every seat for every showing in both theaters *Nanking* played in its two-week New York premiere.

"Why is this so stupid and broken?" he wondered, and proceeded to do the math. Each year, the independent film festival Sundance gets 9,000–10,000 entries. Of them, 124 are accepted as official selections. Of them, eight to 10 are acquired by studios or distributors. Eight to 10 — and even those rarified few are hard-pressed to succeed. Of the 30,000 U.S. movie theaters, only 400 typically show indies. *Nanking* was the rare documentary to approach $1 million in box office. Leonsis estimates it was seen in theaters by about 70,000 people — an audience which nonetheless, in a world where everybody has a movie screen on their computer, or cell phone, struck him as preposterously small. "I wanted to fix this system." As Leonsis' Georgetown University mentor used to counsel him, "The key to success in life is connecting the dots," so that is precisely what he did.

Leonsis also happens to be chairman of Clearspring, probably the largest widget vendor in the country. What if doc and indie films were embedded in widgets and permitted to spread virally, via social networks — especially networks built around charities or social causes? The films could be somewhat monetized by 90 seconds of ads per hour of content, and distributed free for exhibition on an unlimited number of computer or mobile screens. The internet, in other words, would be the largest chain of theaters in the world. Leonsis was particularly intrigued by the philanthropic possibilities. Not only could the movie content focus on issues of social interest, every user who snagged a movie widget from the central website (or from someone else in his or her social network) would be a *pro bono* exhibitor..

"They don't have a lot of time," Leonsis says, "and they don't have a lot of money, but they have a lot of pixels."

Eureka! SnagFilms.com was born. The site launched in July 2008. Less than 10 months later, it had aggregated 25,000 affiliate exhibitors. The ad revenue is split between Snag and the content owners.

Leonsis doesn't imagine his system necessarily monetizing an industry built on friends, family and maxed-out credit cards. He does see it as dramatically expanding the reach of worthy films that otherwise would sink promptly into oblivion. When you have put your heart, soul, years and personal fortune into a film, actual viewership is no small reward. He himself is thinking of *Nanking*, for which he will soon regain distribution rights. It will go up on SnagFilms.com, with a push from Chinese organizations, Christian missionary organizations, anti-war organizations and so on.

"My bet," Leonsis says, "is that a million people will see this film."

Now What Was that You Were Saying, Buddy?

Now then: back to branded fly swatters.

Buddy Media, one of the biggest creators of branded apps, fills two floors of a slightly shabby office building on Broadway just above Columbus Circle. It used to be a Fred Astaire Dance Studio, with a ground-level Indian restaurant and water seeping down the bare-brick walls after every rain. Now it's a code factory, where workers load raw zeroes and ones into their iMacs and forge software parts — parts that are in turn assembled in various combinations to form custom applications. Using that small inventory of a few hundred in-stock parts, Buddy Media can turn out widgets fast and cheap.

"If you're going to do a 728x90 banner ad," says CEO Lazerow, "you might as well do an app. Because it's going to take the same amount of time and cost."

What he doesn't advise is buying for reach and frequency. His buzz term is "reach and engagement," the idea of cultivating a few people instead of pestering a lot more. "Instead of reaching 80 million people, let's reach a million in your target and spend 10 minutes with them." Buddy Media has no difficulty establishing the engagement part. Its hair-do widget for *InStyle* magazine had more than 300,000 installs, 185,000 in the first six weeks. The average time spent on each visit was seven minutes — three hairstyles' worth — and nearly half of the users returned to it more than 25 times. "They basically cost less than traditional banners and you get 75 times greater spent than with regular banners and five times more time spent than with TV ads."

One believer among his clients is Keith S. Levy, vice president of marketing for Anheuser-Busch, which created a Bud Light Dude Test widget to leverage [Note to reader: I have just used the word "leverage" as a verb. This will never happen again] a Bud Light ad called "Dude."

"I think the multiplier effect of the web is extremely powerful," Levy says. Though the 300,000 downloads are laughable compared to a TV buy, "You're really getting a relationship with the consumer." Another widget, created just for the lucky winners of The Bud Light Party Cruise promotion, for instance, created an ongoing community of 4000-some evangelists like ROWYCO®, who (according to his MySpace page) is a 24-year-old Arizonan whose nickname is an obscene acronym, whose slogan is "Hardcore for Life," who likes country and metal and pimped out motorcycles, and who is working on a business degree as Paradise Valley Community College. His friends are Judith, Diana, Courtney (!), Crazy Christene, Justin, The Rouch and — right at the top of the list — Bud Light Party Cruise.

Compare this effort, for instance, to BudTV, which at a cost of $15 million for the first two years alone, attempted to create a content destination more or less paralleling the tube. What Anheuser-Busch earned for its trouble was a squizzilion views of the hilarious commercial "Swear Jar," the enduring enmity of many state attorneys general and zero MySpace friends. One day in late November, BudTV's global web ranking, according to the online analytics site Alexa, was 26,253,061. To put that in perspective, moisttowelettemuseum.com was ranked 5,681,20947. As they say over at A-B, lessons were learned. Though he won't characterize BudTV as a boondoggle, when pressed to look at the relative efficiency of BudTV and pocket-change widgetry, Levy offers, "Did we have to build a stationary network where people have to go and get stuff? No"

Of course, engagement — and even community — cannot be directly correlated to sales. But, excuse me, apart from direct-response advertising, what can? As for the other outstanding issues casting suspicion on the sustainability of widgetry, let's take one more look:

1) Platform incompatibility. While some functionality is sacrificed, something close to universal code is not far off. Lazerow says his apps are, with a minimum of tweaking, one-size fits all.

2) What works for a sexy brand might not work for Charmin? Upon further reflection, why not? Given about two seconds thought, I came up with about 10 toilet-paper-relevant ideas in varying degrees of offensiveness, from SoftCam (rotating video of a basket of kittens, baby butts, ducklings, etc) to a Full-of-Shit-o-Meter (feeding news quotes from celebs, athletes and pols that are transparently disingenuous or worse.

3) Cost. Yes, $5 an install can add up, but many vendors charge much less. More to the point, though, who says that the calendar is the right allocation tool for marketing expenditures? As long as we're reinventing commerce, should we not consider the possibility that marketing programs will be financed for as long as they perform, without arbitrary campaign boundaries? "I guess I can't understand the marketers' narrow-minded definition of the controlled calendar, because that's not how consumers minds work," says Rex Briggs of Marketing Evolution. "It's like Coca Cola

saying, 'All those people with the Coke memorabilia from the '50s, I want it out of their house because that's not the Coke message today.' "

4) Shelf space. Even if you accept that there is a finite amount of real estate on the world's 500,000,000 social-networking pages, the universe is expanding by the second. "Saturate the market?" Lazerow says. "We're not even close. I can't see a world in which we're going to saturate this market."

Oh, and one final thing. If you are a marketer who's spent the past decade investing in a robust website to attract customers and prospects, and you're therefore disinclined to cannibalize your traffic by giving away Website-in-a-Can, don't get too smug. Your audience is making that decision for you.

In a three-month period to begin 2009, according to Alexa, Apple.com's page views per user were down 9%; Comcast.net down 1%; Dell.com, down 22%; AT&T.com, down 18%; Xbox.com, down 9% and so on as corporate e-bastions began to experience the same audience fragmentation that is killing old media. "As popular as your site may be," says Niall Kennedy, the reality is that people are actually visiting Yahoo, MySpace, Google and Facebook thousands of times more than they're visiting you." Tony Zito of MediaForge calls this "the slow death of the destination website." Consider the source — the man sells widgets — but even if he's hyperventilating, the trends are a bit ominous. If Mohammed has indeed cut back on his visits to the mountain, it may be time for the mountain to go to Mohammed.

After all — Ding! — the fares are pretty low.

COMCAST MUST DIE

NEVER PICK A FIGHT," Mark Twain reputedly observed, "with a man who buys his ink by the barrel." It was certainly true. A powerful publisher, if he were irritated enough, had the wherewithal to bury any adversary. It's still true, actually, although, the truism needs to be slightly amended. In the Listenomics Age, never get in a dispute with someone with access to a computer.

Because if he is aggrieved enough, and righteous enough, and persistent enough, and connected enough, he can bury you. Or, at least, make your life miserable for a long, long time. He doesn't need to have a chain of newspapers; all he needs to have, basically, is fingers and rage.

For people with anger issues, the internet is a cathartic godsend and/or lethal weapon. You know the type; those people who might accept the ordinary indignities of life with reasonable equanimity, but become suddenly radicalized when lied to, cheated, bullied or otherwise personally abused. This kind of person will cheerfully invite a shopper with two items to move ahead of him in the checkout aisle, but will pointedly confront the jerk who butts into a movie line. Even if it means a loud squabble. And he won't back down, because it's the principle of the thing. Likewise, he might, upon being denied an Egg McMuffin at 9 a.m. on the preposterous ground that "we're out of muffins," proceed to — how to put this? — seek further information from management, and continue to press the point until someone opens up a new carton of muffins and fills his order (with God knows what employee DNA secreted inside.) This sort of guy might hypothetically even be disinvited from a transatlantic flight, on security grounds, were an airline counter employee to mishandle visa documentation, then lie to his face about seating availability, then, caught in the lie, choose to regard his resulting swearword and general seething as a sign of imminent violence and bounce the poor SOB from the flight. And the next one.

In other words, a guy just like me.

Some of us — no matter how generally sweet-natured and generous, no matter how friendly and thoughtful, no matter how empathetic and transcendently kind — are simply not to be fucked with. Because when we are wronged, we will go to rather extreme lengths to be righted. Back in September 2007, this was something Comcast Corp. did not know.

I have since read thousands of consumer horror stories about that company, so I can safely say my particular experience was more or less ordinary, and I will therefore provide the bare minimum of detail. But experience with Comcast customer service creates the tendency to define deviancy down. For you, the uninitiated, reading about this episode may still cause you to bleed through the ears.

Really, Don't Read This. It's Unhealthy.

The saga actually began the previous July when I tried to order the "Triple Play." This is a bundled package of cable, VOIP phone service and broadband for a low, low promotional price. I was sick of dealing with three providers, and I anyway needed a technician to connect a cable to a computer situated in a dead zone in my house, where my wireless router frequently lost its connection. So I decided to throw caution to the wind, put all my eggs in one basket and — wait, one more cliché — prove I didn't have the brains I was born with. I sat on the phone with a Comcast sales rep for 15 minutes until he had all the information he needed. He answered all my questions to my satisfaction, including the matter of my hardwiring needs, to which he voiced no objection. Then he informed me his system was down, and he would call me back later to finish the transaction. He never did.

This is what you call "an omen."

Two months later, failing somehow to register God's heavy-handed foreshadowing, I started the whole process over. In retrospect, this seems like an act of extraordinary recklessness, like flying a Cessna into a thunderstorm. What can I say? I thought things would clear up. Tragically, I strapped myself in and took off, on a collision course with destiny — or, at least, corporate indifference on an epic scale. The faint-hearted can skip the next 11 paragraphs. Here it comes:

In the late summer of 2007, having ported my phone number from my previous carrier, I made an appointment for installation, on a Sunday in the 11 a.m. – 2 p.m. window. At 1 p.m. on the appointed day, the installer called to say he was finishing another job and would be a little late. By "a little," evidently he meant "3 ½ hours." At 5:15 p.m. I left to take my daughter to college. At 5:30 p.m. he apparently showed up to my house, found me absent and cancelled my entire installation. During this time, naturally I called Comcast for information. Three calls yielded 40 total minutes on hold, followed by three separate customer-service agents promising to call me back. None ever did. Finally, I was informed that when the tardy installer cancelled my job ticket, he also negated my porting authority to switch my old phone number to a Comcast. In other words, I had to start all over again. Again.

This was no omen. It was a freakin' air raid siren warning me to escape for my life. Nope, not me. I like to finish what I start. The installation was rescheduled for two weeks hence. Sure enough, two weeks later, an installer arrived. The first thing he told me is that I can't have a second computer hardwired to the cable modem; I have to have a second account. That was news to me, and had I known it when I signed up, I wouldn't have bothered. But the sales agent who took my order, although I informed him of my hardwiring needs, never bothered to mention anything about one-cable-per-account. Once again, now for the third time, this misunderstanding augured poorly. Any clear-thinking mortal would have sent the guy packing, but 1) I'd already cancelled my satellite TV service, and 2) I am a fool.

The fellow came in and got busy. Very busy. The installation was taking longer than he planned, including some wall drilling he hadn't anticipated, because, after all, it's not as though he makes a living threading cable into houses. He said he had to run out to another installer on a nearby job to borrow a drill bit. Five and a half hours later, he hadn't returned.

Two of four telephones in the house did not work, and neither did a TV.

Naturally, I called my friends at Comcast customer service — the ones who had kept me on hold for 40 minutes two weeks earlier while busy not locating my installer. Let me just briefly emphasize something: 40 minutes on hold. You can do a lot in 40 minutes. You can drive from Washington to Baltimore. You can play two games of Candyland with your six-year-old. You can cook a steak then watch an episode of *Seinfeld*. You can perform a very respectable heart catheterization. Or you can languish on hold with a cable company as your blood pressure steadily rises.

Jihad to be There

This time, mercifully, the initial phone queue was brief. I explained my circumstances to the customer service rep requested, who requested a "a quick moment" to investigate. Fifteen minutes later, she returned to ask for "one more moment." Then I was left on hold for 32 minutes. During that time, I used my cell phone to call customer service and ask for a supervisor. I was not permitted to speak to one. I was told somebody would call me back. I thanked them and went back to the first call. In "quick-minute" number 50, she resumed the line to inform me "a tech is on the way." I told her I have been on hold for a total of 48 minutes. She said "I doubt that." I asked for a supervisor. She told me she couldn't connect me; someone would call back. Nobody did. In the meantime, I missed a 4:30 p.m. appointment, but had a dinner commitment at 6:30 p.m. I could not cancel. At 5:40 p.m., the installer reappeared, none too sheepish about his absence.

Then, a miracle occurred: Simultaneous with his reappearance, Comcast called me back, informing me that the installer had lied to his dispatcher and reported my installation complete. We agree that we are both of us victims — albeit some of us more than others. As he had been AWOL for 4 hours and 10 minutes and had but 34 minutes to finish the job, I asked to have the remaining work rescheduled for the next day. OK, not quite requested. More like demanded. More like loudly demanded. More like I was in a bank with a sawed-off shotgun, and she wasn't emptying the cash drawer quite briskly enough to suit me. No matter. Notwithstanding my rage and gathering apoplexy, my request was beyond the scope of her authority. All she could do was confirm the liar was back at my house. At this point — though I'd actually become afflicted with tachycardia, and my heart raced in my chest — I ceased shouting and commenced some spirited begging. It wasn't dignified, but it seemed like the thing to do. "Please," I whined. "Puhlleeeeeeasssse!" Finally she relented. I was briefly on hold when someone named Carol Webb got on the line.

Then, another surprise. She quickly became the first Comcast employee in the entire day of madness to offer an apology. "I am dumbfounded by what has happened to you," she said. The sentiment was appreciated, but, myself, I was at that point far beyond stunned silence. What's the opposite of dumbfounded? I was suddenly feeling quite loquacious. But I didn't share my thoughts with Ms. Webb. I shared them with *the* web.

It was just a brief post on my on my AdAge.com blog. In it, I summarized the events recounted above and added a few personal thoughts:

"Is this company so frantic to seize market share on voice and broadband that it is willing to disrupt customers' lives, fail to appear, repeatedly lie to them, walk out on them and then treat the customer as if he or she is a nuisance? Well, we shall see. This is the Listenomics age. We will not take it quietly."

Probably more notable than the text of the rant, though, was its headline — which, in my view, had a nice ring to it: "COMCAST MUST DIE: Seeking Ideas for a Consumer Jihad."

There is something you need to know about my *Ad Age* blog. Nobody reads it. Anyway, hardly anybody. But somehow people read this one. And somehow I had struck a nerve; within 24 hours, there were 60 comments, which compared very favorably to my previous average of 0. Some were longwinded recitations of horror stories far worse than mine. Some were words of encouragement and support for a Holy War against Comcast and the whole Co-axis of Evil. Yet others wrote in to give me exactly the help I solicited. Within an hour, someone called my attention to the YouTube video of a cable installer asleep on the customer's sofa, where he had lapsed into slumber while on hold with his own company. And another had this to say:

"A consumer Jihad is exactly what is needed. Perhaps what should be done is to buy the web domain ComcastMustDie.com and establish a blog so that all of the folks that have been screwed by Comcast can tell their story. You could put a link to the new Comcast Jihad website and release a press release to drive traffic to the site. The anger and disgust expressed there would be huge."

"Blackmail" is Such an Unpleasant Word

Hmm. Using the web to galvanizing consumer anger and disgust. An interesting idea. Of course, as I well knew, it had been done before. In November 2004, an English gent named Adrian Melrose bought a brand new Discovery 3. It was a $60,000 lemon. But his dealer, and Land Rover itself, were insufficiently responsive. So he began a blog called "Discover the Truth About Land Rover Discovery 3" and took his complaint public. For weeks, Land Rover ignored the blog even as Melrose began to accrue hundreds and, eventually, thousands of sympathetic comments. The notoreity eventually became so embarrassing they replaced the man's car — whereupon the replacement immediately broke down, too. Then, while they were fixing that, a loaner broke down. Then, a second loaner broke down — all of this misadventure being chronicled, of course, on Melrose's blog.

Finally, they refunded him his money, triggering accusations that he had blackmailed — or, shall we say, blogmailed — the company into submission. But it wasn't blackmail, at all. Melrose was simply exerting the leverage that the digital age has bestowed upon the consumer to make lemonade out of lemons. What the makers of Discovery discovered is to take care of whom you screw, lest you turn into the screwee. And it's no one-night stand, either. To this day, if you Google that model, you encounter Melrose's blog on the first page — at God knows what cost to the company in showroom traffic. Yet, incredibly, the brand still has no online forum for FAQs and service bulletins, much less complaints. Melrose offered to turn over his blog for that very purpose, but the company declined. As he put it in the summer of 2007:

> Land Rover U.K. needs an efficient platform to listen and care for their customers. The reason I am getting 700 hits to www.haveyoursay.com, mostly search engine driven, is because their existing customers want to talk to one of there favourite brands — just like me — they want to be loyal — but they are frustrated 'cause they think nobody listens. They have no way of communicating with Land Rover save through the Royal Mail, and that doesn't work for them.

Adrian Melrose achieved Google perpetuity more or less by accident. In what may have been a defining moment in Listenomics, someone else contrived to the same thing on purpose. This was on June 21, 2005, when blogger Jeff Jarvis, of Buzzmachine.com, finally got fed up with Dell, the computer seller.

"I bought a Dell laptop and it didn't frigging work," he says, but that was just the first part of the problem. The second part was Dell's handling of his complaint — which, for want of a better term, was sadly Comcastic. "Too many emails, too many phone calls, too much frustration. I went on my blog and I created a post headlined: 'Dell Lies, Dell Sucks.' That wasn't just an indication of my immaturity. It was search engine optimization."

Yes, Jarvis knew that his blog was so widely read, and more importantly, widely linked to, that any subsequent Google search for "Dell" would yield a results page prominently displaying "Dell Sucks." In fact, the last line of his post was "Put that in your Google and smoke it, Dell."

Eventually, like Land Rover and Comcast, Dell upper management intervened and Jarvis got his refund, but not before the situation got far beyond their control. The saga was reported on by the *Houston Chronicle*, the (U.K.) *Guardian* and *BusinessWeek* magazine, among others, which fed web chatter, which resulted in more mainstream coverage and so on. Very soon Dell was the most infamously sucky company on earth. Now, if you type "Dell sucks" into a search engine, you will get back 2 million results. One man had tapped a deep vein of consumer frustration — frustration that in the internet age now can rise, like a geyser, to the surface. As Jarvis puts it, "If you go online and type in your search engine '(any brand) sucks,' you will find the real *Consumer Reports*." Yet for a long time, Dell ignored the controversy, a lapse that cost the company mightily in public image and most likely the bottom line. Correlation isn't causation, but it's worth noting that 17 months after Jarvis's post, the company lost its world leadership in computer sales to HP.

Bitch, Bitch, Bitch

Jeff Jarvis, of course, no more invented online bitching than I did. From its earliest days, the internet has been a breeding ground for the culture of complaint. Technorati, the blog search engine, claims to scan 109 million blogs. On October 27, 2007, it tracked 1,232,853 posts using the word "sucks." On Blogdigger: 234,448. On BlogPulse: 641,682. And on Google Blog Search: 3,264,834.

That is a lot of sucking.

Among the items of suckitude: Bill O'Reilly; PayPal, school, skateboarding, milk, "Survivor," cancer, Facebook, everybody, Garfield (no relation), and — hilariously enough — the Morphy Richards Pod Bagless Compact vacuum cleaner. (Although, you know, not in a good way.)

That's from about 50 of the first few entries. I can't speak for most of the other 3,264,784 because very quickly all that whining gets a little tedious — although I couldn't help but notice that one blogger, demonstrating perhaps the world's highest threshold of satisfaction, found fault with orgasm. That is a person whom Bill O'Reilly, PayPal and Facebook might find difficult to please. (By the way, while 3,264,784 bloggers were pissed off at that particular moment, many others were not. And they blog, too. Google reckoned that 525,240 posts that day invoked the word "awesome." Another 21,734 described "an incredible experience." And 35,060 deemed something or other "the coolest thing ever." Coolness, though, is also in the eye of the beholder. Exactly one blogger, a college student named Suze Bramlet, used "coolest thing ever" in the same thought with "Birkenstocks," but otherwise the pool of cool is predictably deep: Librarything.com, Google Earth, Google Ride Finder, BenGay, Pandora, the Diet Coke and Mentos video, tennis and the fact that nurse sharks have two uteri but no placenta. So, if you're depressed about what consumers think of you, cheer up. Just be as cool as a nurse shark with two uteri but no placenta.)

Obviously, whiners are a category of consumer that has always existed. There have always been chronic malcontents out there, in addition to the loudly, legitimately aggrieved. Till recently, however, their audience has been limited to the offending party and a few unfortunate intimates. Now, thanks to the internet, their audience is potentially everyone. And therefore, nobody anymore need depend on critical judgments from astonishingly perspicacious elites such as myself. Opinions are like parathyroid glands; everybody has one. About politics, about sports, about mouthwash.

> *"Listerine was chugging along nicely from its introduction as the first over-the-counter mouthwash in 1914, killing germs and tasting like shit until 1992,"* blogged a fellow named Stegmann. *"This is when Cool Mint Listerine was introduced, probably to combat that sickly-sweet but non-antiseptic upstart Scope. Since then there have been several new flavors of Listerine introduced, including Natural Citrus and Cinnamon.... The latest variety of Listerine is Vanilla Mint. It's*

*advertised as being 'less intense.' Does it taste good? Sure, I'm drinking
a glass of it right now, poured generously over ice. Yum. Listen ... I don't
want my Listerine to be delicious. I want it to taste horrible and kill
germs, just like it did 100 years ago. What sissies we've all become."*

God bless Mr. Stegmann, and duly noted. We'll mark him down as
against Vanilla Mint Listerine. What makes his post remarkable and
revolutionary, though, isn't that he thinks what he thinks. It's that he
took the trouble to inform the world at large, and that at least one member
of the world at large took the trouble to listen. Multiply that times
a billion and you grasp the Listenomics world. There are ever-growing
numbers of mechanisms for expressing opinions, and ever-growing
mechanisms for seeking them, some more insightful and eloquent than
others. The following is a Netflix user review of *The Prince and Me*:

*The improper use of grammer clearly stears me into the direction that
the movie isnt worth the watch and I was correct .The Prince and Me
more along the lines of The Prince and I. When will people learn? If you
cant even get the title right. Why waste the effort and watch it???*

OK, that one individual disappointed viewer happens to be a numb-
skull. Yet for every orthographically challenged apostrophe-deprived
(and, incidentally, mistaken) grammar stickler, there is also someone
who understands the art of complaint, and how to get an audience.
The following letter was sent to the British cable company NTL taking
issue — as it happens — with precisely what infuriated me about Comcast. It is what diplomats call "a frank exchange of ideas," so I warn you
on language grounds. It's also about three pages long, but I reprint the
whole thing without apology. It's a classic that eventually became an
internet viral:

Dear Cretins,

*I have been an NTL customer since 9th July 2001, when I signed up
for your 3-in-one deal for cable TV, cable modem, and telephone.
During this three-month period I have encountered inadequacy of
service which I had not previously considered possible, as well as*

ignorance and stupidity of monolithic proportions. Please allow me to provide specific details, so that you can either pursue your professional prerogative, and seek to rectify these difficulties — or more likely (I suspect) so that you can have some entertaining reading material as you while away the working day smoking B&H and drinking vendor-coffee on the bog in your office.

My initial installation was cancelled without warning, resulting in my spending an entire Saturday sitting on my fat arse waiting for your technician to arrive. When he did not arrive, I spent a further 57 minutes listening to your infuriating hold music, and the even more annoying Scottish robot woman telling me to look at your helpful website.... HOW? I alleviated the boredom by playing with my testicles for a few minutes — an activity at which you are no-doubt both familiar and highly adept The rescheduled installation then took place some two weeks later, although the technician did forget to bring a number of vital tools — such as a drill-bit, and his cerebrum. Two weeks later, my cable modem had still not arrived.

After 15 telephone calls over 4 weeks my modem arrived ... six weeks after I had requested it, and begun to pay for it. I estimate your internet server's downtime is roughly 35% ... hours between about 6pm -midnight, Mon-Fri, and most of the weekend. I am still waiting for my telephone connection. I have made 9 calls on my mobile to your no-help line, and have been unhelpfully transferred to a variety of disinterested individuals, who are it seems also highly skilled bollock jugglers.

I have been informed that a telephone line is available (and someone will call me back); that no telephone line is available (and someone will call me back); that I will be transferred to someone who knows whether or not a telephone line is available (and then been cut off); that I will be transferred to someone (and then been redirected to an answer machine informing me that your office is closed); that I will be transferred to someone and then been redirected to the irritating Scottish robot woman ... and several other variations on this theme.

Doubtless you are no longer reading this letter, as you have at least a thousand other dissatisfied customers to ignore, and also another one of those crucially important testicle-moments to attend to. Frankly I don't care, it's far more satisfying as a customer to voice my frustration's in print than to shout them at your unending hold music. Forgive me, therefore, if I continue.

I thought BT were shit, that they had attained the holy piss-pot of godawful customer relations, that no-one, anywhere, ever, could be more disinterested, less helpful or more obstructive to delivering service to their customers. That's why I chose NTL, and because, well, there isn't anyone else is there? How surprised I therefore was, when I discovered to my considerable dissatisfaction and disappointment what a useless shower of bastards you truly are. You are sputum-filled pieces of distended rectum incompetents of the highest order.

British Telecom — wankers though they are — shine like brilliant beacons of success, in the filthy puss-filled mire of your seemingly limitless inadequacy. Suffice to say that I have now given up on my futile and foolhardy quest to receive any kind of service from you. I suggest that you cease any potential future attempts to extort payment from me for the services which you have so pointedly and catastrophically failed to deliver — any such activity will be greeted initially with hilarity and disbelief quickly be replaced by derision, and even perhaps bemused rage.

I enclose two small deposits, selected with great care from my cats litter tray, as an expression of my utter and complete contempt for both you and your pointless company. I sincerely hope that they have not become desiccated during transit — they were satisfyingly moist at the time of posting, and I would feel considerable disappointment if you did not experience both their rich aroma and delicate texture. Consider them the very embodiment of my feelings towards NTL, and its worthless employees.

Have a nice day — may it be the last in you miserable short life, you irritatingly incompetent and infuriatingly unhelpful bunch of [this expletive definitely deleted.]

Yours psychotically,

— John

Cancel the Account. CANCEL THE ACCOUNT.

John may or may not have been driven insane by his cable company, but he was decidedly an angry loner, one who just happened to find a larger audience. Others are a lot more organized. Having reckoned that consumer appetite for information and the impulse to venture opinions are both bottomless wells, a number of businesses have undertaken to host the give and take. Amazon.com and eBay both solicit user ratings, of books and products but also of the sellers who trade on their sites. Angieslist.com solicits ratings of local plumbers, roofers and other contractors and home services, and charges members a monthly fee for access. Yelp.com and citysearch.com are advertising-supported sites for rating restaurants, stores, nightlife and local media community by community. ePinions.com does the same on a national level, mainly for national brands. . These sites have an impact.

Craig Stoll, owner of San Francisco's Delfina and Pizzeria Delfina, found his restaurants doing very well by traditional critics, but intermittently getting slammed on Yelp. He didn't necessarily see a drop in business, but it was tough on staff morale. And it just felt so unfair. So, as an expression of defiance, and frustration, his wait staff started wearing t-shirts emblazoned with excerpts from the reviews. Not the *good* reviews, the savage ones. To wit: "The pizza was sooooo greasy. I am assuming this was in part due to the pig fat."

Stoll acknowledges that some of the Yelping has helped him identify service and menu issues he'd been unaware of, but the powerlessness to control random sniping just drives him crazy. "I stopped going onto Yelp a couple of years ago," he says. "I'm just — I can't take it. You know, it just puts me in a bad mood for the entire day." On the other hand, he wryly observes, "The other day one of our managers was deciding on where she was going to go [to dine], and was looking at Yelp."

I'm skimping on the detail here, but let's face it: the novelty attached to consumer ratings, compared to, say, the Jurassic period of 2001, has long since evaporated. These mechanisms are now simply a part of the landscape — so much so that they've long since become a target for parody. In 2006, *The New York Times* discovered that hundreds of people had posted Amazon.com customer reviews of a gallon jug of whole milk. Among the comments sifted out by the *Times*: "I give this Tuscan milk four stars simply because I found the consistency a little too 'milk-like' for my tastes." Another advised, "One word of caution — milk, even when frozen into a baseball-bat shape, is nigh worthless as a baseball bat, merely shattering into cloudy fragments at the first strike of a baseball."

Okay, granted, the Willy Loman-esque craving for attention does sometimes verge on the absurd, but it can also yield some remarkable stories. After all, what is drama but conflict? The confrontation of Big Powerful Forces by righteous Everyman can be quite riveting, quite satisfying and even sometimes quite funny. This is the premise behind Consumerist.com, a gossipy, ad-supported site in the Gawker blog empire, founded in 2005 as a forum for exposing substandard goods and services in the Gawker tradition of persistence and snark. "Basically," says editor Ben Popken, "to inform, empower and entertain consumers." This is where a fellow named Vincent Ferrari became a national icon by doing nothing more than trying to cancel his AOL account — and recording the phone conversation. The Consumerist posted the audio, as the AOL employee worked his way down the customer-retention flowchart to the point of hilarious obstructionism. An excerpt:

> **AOL:** *Is there a problem with the software itself?*
> **Vincent:** *No. I just don't use it. I don't need it. I don't want it. I don't need it anymore.*
> **AOL:** *So when you use the computer, is that for business, or …?*
> **Vincent:** *Dude, what difference does it make? I don't want the AOL account anymore. Cancel it.*
> **AOL:** *Last month was 545 hours of usage.*
> **Vincent:** *"I don't know how I can make this any clearer, so I'm just going to say it one more time: cancel the account"*

AOL: What'sa matter, man? I'm just trying to help you here.

Vincent: *You're not helping me. I'm calling to cancel my account. Helping me would be canceling the account.*

AOL: No. It wouldn't actually.

Vincent: *Cancel the account. Cancel the account. CAN-CEL THE AC-COUNT.*

And so on. As a result of the Ferrari tape, AOL changed its intervention procedures to make it easier for customers to cancel. They also fired the employee, loyal and dedicated though he was. That's what happens when consumerism becomes a spectator sport.

The Consumerist mounted another mini-crusade against Wal-Mart, when it discovered that the world's largest retailer was selling t-shirts emblazoned with the skull-and-crossbones insignia of the Nazi 3rd SS Division, the so-called Totenkopf division dedicated mainly to guarding concentration camps. The story surfaced on another blog, but The Consumerist flogged it for all it was worth — even after Wal-Mart promised to pull the merchandise. Because, as it turned out, long afterwards people kept finding the t-shirts at Wal-Mart.

"We have infinite pixels to spend," says Ben Popken, "so we were able to follow it. Day 5, Day 27, Day 97 of the Nazi-recall watch. We can follow it along to doomsday." Just as they did with a line of sketchy Wal-Mart flip-flops, which were so laden with formaldehyde that they scarred the feet of people who wore them. Wal-Mart at first denied a problem, but the website eventually wore them down.

"It levels the playing field, it swings the pendulum back in the consumer's direction. Now one person has the same power as a mega corporation. There's no longer such tight control over messages, by entrenched, hierarchical powers. Everyone has the same power to be heard. The determinant of who gets heard is not who has the most media dollars, but has the most interesting thing to say."

Yet one more player in the burgeoning e-grievance industry is Complaints.com, which has less of an audience than The Consumerist, but a bolder promise: to explicitly shame the company you don't like. To be specific, the site boasts of employing the Jeff Jarvis strategy of search-engine optimization. As the site's homepage proclaims: "Often, a single complaint posted to Complaints.com about a business appears higher in the search result rankings than the home page of the business that is the subject of the complaint."

Funny. That was just my plan for Comcast.

Or, as I Childishly Called it, "Qualmcast"

Once again, though, AdAge.com/Garfield — a.k.a. the Bobos-phere — is no Buzzmachine. It is a venue mainly for writing this very book, where, according to the principles of Listenomics, I would post a few passages here and a few there for the consideration of my readers, who were free to comment, correct, argue and amplify. The process was often quite productive, but seldom exactly a free for all. My daily page views were in the dozens or the hundreds, but by no means the thousands. Until I commenced a Holy War against my cable company. That's when things started jumping.

In the Chapter 4, I discussed "seeding," a mechanism through which your would-be viral is thrust to the attention of presumably friendly parties via public relations or email. *Ad Age* helped me there by including my post in our daily email blast to 175,000 of our 700,000-some registered users. Suddenly, I had 60 comments, most fulminating with Comcast hatred, and soon other bloggers — notably Jeff Jarvis himself — were helping me get the word out. So I kept plugging away at Comcast Must Die: Parts II, III, etc. This was my third post, from Sept 14, 2007, headlined: "The Fix is In."

Craziest thing. Shortly after being held up to ridicule, contempt, anger and a fair amount of loathing, Qualmcast was all over itself to finish my aborted install and attend to all of my service grievances large and small. Two extraordinarily pleasant and helpful supervisors were in my home for four solid hours. I got four calls from customer-service people following up on my "issues," and I'm told I have a phone message wait-

ing for me at home from a Qualmcast VP. No doubt it is simply ooz-ing with regret. And that regret is sincere. Qualmcast seriously regrets that the customer they mistreated so brutally was me. Because I have an audience, and friends in the blogosphere, including Jeff Jarvis, who has helped me spread the infuriating story far and wide. Qualmcast senior director of corporate communications Jenni Moyer deemed the outbreak of hostility as something that must be contained, so she issued a press release:

"We are appalled by the experiences that some customers have recently shared on blogs and in other forums. Where we have been able to iden-tify customers who have had unsatisfactory service interactions, we have taken action to fix their problems. We recognize that it should not take a public event to have good customer service, and we are work-ing hard to ensure that all of our customers receive the best possible service."

Not too bad. She didn't actually lie till the last clause of the last sen-tence. As the comment traffic has made abundantly clear, Qualmcast is not working hard to ensure that its customers are receiving the best possible service. It is working hard to reduce costs to be competitive with the other telecoms, who also treat customers shabbily, in order to compete with Qualmcast.

But the other stuff was true. They have taken actions to correct my problems, which may be lucky for me, but in no way pacifying. On the contrary, as a jihadist bent on destroying their corrupt system, I am angrier and more zealous than ever. Customer service doesn't mean kissing the ass of VIPs and putting everybody else in the hold queue till Groundhog Day. It means treating all your customers with dignity and respect, and investing all necessary resources to see that problems get solved immediately—for everyone. Which, if institutionalized in company culture, would eventually cease to be an expense and instead be a priceless differentiator in a commodity category.

Think about Nordstrom. It has higher costs than its competitors, but also commands a price premium, because, duh, customers attach value to being treated like human beings.

So thanks for all the follow-up, Qualmcast. However, I cannot be bribed with ex-post-facto attentiveness, and I am by no means finished with you. Oh, and by the way, my fucking phones still don't work

It was true. Comcast had indeed put the full-court press on my particular problem. Shortly after that item was posted, there were five Comcast vans parked by the cable hub nearest my house for something like 18 hours. (And now my phone-cable-broadband service is just exquisite, thank you very much). But this hardly pacified me. As you can see, it just made me madder. Their sudden attentiveness implied that my jihad was just "blogmail," an extortion attempt to get my phones attended to. But it was no such thing. Comcast Must Die was an extortion attempt to get *everybody's* phones attended to.

Things were moving along splendidly, with Comcast Must Die getting the attention of bloggers and blogees alike and comment traffic in the Bobosphere picking up apace. But then came three incredible strokes of good fortune, pretty much back to back. You'll recall that on Day 1, somebody suggested reserving the domain name Comcastmustdie.com. By Day 4, somebody had done just that. His name is Bart Wilson. He is founder of a company called Voyager International in Santa Fe, NM, and he was fed up with Comcast, too. Then it was but for me to find someone to actually create the site, and a dedicated blog linked to it. No problem. I've dealt with Comcast; I know how to beg. Bart cheerfully agreed to build-out the back end and two weeks into my jihad, Comcastmustdie.com went live with the following manifesto:

Actually, I have no death wish for Comcast or any other gigantic, blundering, greedy, arrogant corporate monstrosity, What I do have is the earnest desire for such companies to change their ways. This site offers an opportunity — for you to vent your grievances (civilly, please) and for Comcast to pay close attention.

I advise you to include your customer number in your post; this will give Comcast the chance to contact you and work on your problem. If it does so, I encourage you to post an update, giving credit where credit

*is due. Meantime, be aware you may be the target of online phish-
ers trying to get personal information from you. DO NOT REPLY TO
EMAILS CLAIMING TO BE FROM COMCAST. Deal with them only
by phone.*

*Congratulations. You are no longer just an angry, mistreated customer.
Nor, I hope, are you just part of an e-mob. But you are a revolutionary,
wresting control from the oligarchs, and claiming it for the consumer.
Your power is enormous. Use it wisely.*

In the first 24 hours, there were 70 comments. Within a week, there
were 200 — not site visits, not page views. Comments. People taking the
trouble to move from mouse to keyboard and leave their thoughts, most
of them, predictably, excoriating Comcast. Some even came from Comcast employees, as miserable at their workplace as we customers were on
the other end of the cable. A few other Comcasters defended their company, in some cases heaping ridicule on stupid and/or hostile customers.
Comcast PR executives even posted a few official comments, declaring
themselves sensitive to consumer complaints and heavily invested in
making their already dedicated efforts more robust still. Those assertions were met with some skepticism, including many famous-name
vulgarities.

Oddly, I knew that the claim was at least partly true. That's
because very early on in my Comcast nightmare, I received a phone
call from a young woman named Michelle doing what seemed to be
a random customer-satisfaction survey on my "Comcast Triple Play"
installation.

"On a scale of 1 to 10," she asked, "with 10 being best, how would you
rate your service?"

"Zero," I said. There was a prolonged pause.

"May I ask what the problem was?" she continued, and I explained.
She apologized and offered to connect me to someone who could
help. Why not? I was curious if the routine follow-up would yield any
results.

"You will be on hold," she explained, "and it will be silent on the line for a moment."

"OK," I said. "Thank you." What followed was 20 seconds of silence. Then, a rapid busy signal. Comcast, apparently for reasons no more sinister than its utter incompetence, had hung up on me.

The Hammer Lady

I wondered if this was yet another omen, maybe telling me to quit while I was ahead, But then came lucky break number two, a news event so delicious and unexpected that when I heard about it I momentarily suspended my air of preternatural *sangfroid* and cackled maniacally while dancing an improvised jig. I refer, of course, to the criminal behavior of one Mona Shaw. On August 17, 2007, having been subjected to a typically outrageous chain of Comcast abuse and neglect over four days, Ms. Shaw sat at a bench outside a suburban Virginia customer-service center where she'd be sent to wait to speak to a manager. After two hours in the baking sun, she was informed, sorry, the manager had left for weekend. This somehow got under her skin.

"I think they got so bloody big," she told me afterward, "they thought they were completely immune to everyone, they could do anything they damn well please."

The indignity, heaped upon the previous week's worth of indignities, rankled her all evening, and all day Saturday and Sunday. On August 20, she went back to the service center with her husband and a means, she later said, of not being kept waiting. The means were: a claw hammer. This she used to pulverize a computer and two telephones, at which point she wryly remarked, "Have I got your attention now?" She did, indeed. But she knew that already. This was a premeditated attack committed not for the hatred of consumer electronics but for the righteousness of indignation, on the theory, as she put it, "If I can't have phone, okay, let's go after yours." The police had a slightly different legal theory, and immediately arrested her — although almost as quickly the handcuffs were removed so she could be attended to by ambulance personnel. Ms. Shaw has a heart condition and her blood pressure following the incident spiked out of control.

No wonder. She was 75.

Mona thus became the Barbara Fritchie of cable rage. To paraphrase John Greenleaf Whittier: "Bust if you must this old gray head, but fix my fuckin' phone," she said.

Next thing you know, she was featured in the *Washington Post*, on *Good Morning America*, and all over the media all over the world. People sent her money to pay her fine. (She donated the cash to the SPCA). She received, via parcel post, three hammers. And wherever Mona Shaw was reported on, so was Comcastmustdie.com. As I was saying: a lucky break. By November 1, there were more than 900 comments on the site, including plenty an infuriating tale of woe, quite a few employee *mea culpa*s, some union organizing and some corporate-PR boilerplate about how hard the company is working on behalf of all 25 million of its customers, blah, blah, blah. Indeed, between Mona and Comcastmustdie.com, the company found itself answering press inquiries left and right — but few more dishonestly than the one from a colleague at *Ad Age*.

"We treat every interaction the same, independent of Bob's blog or anybody else's blog," said Jennifer Khoury, a Comcast spokeswoman. "Bob didn't want to be treated any differently, and he hasn't." Yeah, sure, simply everybody gets the five-repair-truck/five VP-follow-up-calls treatment. And therein stroke of luck No. 3. In a breathtaking succession of PR calamities over two weeks, the company almost seemed to intentionally make itself a target of consumers, regulators and financiers — first by being ham-fisted with The Hammer Lady, then by blocking users attempting to employ BitTorrent file-trading software, then by preventing Big 10 games from being aired in the Midwest, then by pulling MSNBC off of basic cable in Oregon. Every one of these displays of tone deafness and arrogance, of course, drove traffic to the site. But the most satisfying of that traffic to Comcastmustdie.com was from Comcast itself. Not "independent of" the blog but entirely dependent on it, the company was reading every single entry and, as I'd proposed, following up on every one it could. For example:

> *Update on my October 12 posting at 9:49AM ... This site is fantastic. Quickly after making my post I received 3 phone calls from Comcast: (1) a fellow named Mark called from corporate, left his number and told me I would be contacted by someone from my local office, (2) a call from Gwen who was at the local office letting me know who specifically would be handling my case, and finally (3) Rebecca who was handling my case. After some phone tag (due to my schedule — not Comcast's fault), I was able to connect with Rebecca today who had gone through my bill, corrected all of the charges and let me know my new monthly balance. She also made sure that I was credited for past charges and called to let me know how much my new statement amount was so that I wouldn't overpay"*

There were quite a few of these comments, validating my premise that an outside entity was doing for Comcast what Comcast itself should have been doing all along. Once again, that's also what Buzzmachine's Jeff Jarvis has been *saying* all along. "There's a conversation going on about your brand in the open. You can either join it or not." Could it be that Comcast was taking its first baby steps toward redemption? Others have.

Consider the suddenly reflective General Motors, for decades the quintessence of corporate arrogance. Its then-vice chairman Bob Lutz now hosted a blog, sprinkled with comments from those who despise GM products and say so — among other aspersions, such as accusations of a worldwide conspiracy to suppress electric-car technology:

> *GM's "Volt" PHEV (Plug-in Hybrid Electric Vehicle) is the traditional auto industry's latest attempt to misinform the public about the viability of EVs (Electric Vehicles) and PHEVs. Most corporate media outlets covered the story and repeated GM's propaganda that the battery technology is still not ready and that it will be very expensive. All the media outlets also specifically cited the Volt's limited 40 mile electric-only range. This is also GM's attempt to implant in the public mind the*

idea that EVs and PHEVs only have a 40 mile range. The truth is that the extremely reliable heavy-duty NiMH battery technology designed by Stanford Ovshinsky is ten years old and was bought by GM and then sold to Chevron, who is now sequestering the battery and refuses to sell it to small start-up EV manufactures…

Why expose yourself to such flak? Because it's being fired up all around you anyway. Lutz believed it's better to have the conversation on your own turf than behind your back. Better to harvest valuable insights about your products and brands. Better to be able to influence perceptions, than to be a helpless bystander.

(No doubt the same rationale drove GM's handling of the consumer-generated-ad fiasco, in which as part of a consumer promotion it posted audio and visual elements for a Chevy Tahoe commercial, and asked consumers to assemble them, and embellish them, into a finished spot. Oh, consumers did — populating Chevy's contest site and YouTube with endless iterations of one idea: that the gas-guzzling Tahoe was a grotesque offense against a green earth. To the company's eternal credit, it did not remove those entries — on the theory, correct, that hosting the conversation, however antagonistic, would win more goodwill than clumsy corporate censorship.)

Yes, sometimes, miraculously, the scales fall from corporate eyes.

Hi-Ho the Derry-O, The Blogger in the Dell

For instance, if you were to ask Jeff Jarvis today, he would tell you that, no matter what you might have read online, Dell doesn't suck. In fact, he told his readers two years after the episode, his erstwhile nemesis has catapulted "from worst to first."

They have achieved this with a multi-pronged Web 2.0 assault on the status quo. It began with a corporate blog, Direct2Dell, which invited all comers to weigh in on all matters — including suckitude. This came in handy when Dell laptop batteries started to spontaneously combust like a Spinal Tap drummer. Here the company also learned, in bloodcurdling detail, just how inefficient was their customer service infrastructure.

By focusing on (apparent) cost, the managers outsourced to call centers too far and wide and empowered too few employees to divert from their scripts to actually solve a customer problem. The actual cost of the system revealed itself after Jarvis lost his temper, and Dell lost market share. And market value; once again granting that no correlation can be proven, in the year that followed, Dell's share price was carved in half. While it's possible to ignore your Google results, it is impossible to ignore Wall Street.

The tiger's new stripes all culminated in Dell IdeaStorm.com, an online repository for customer suggestions and interactions about all things computing. It is part complaint box, part suggestion box, part polling place and part social network and it has transformed the culture of the company from blind cost-cutting to dialogue. Of course it's impossible to really know anybody's motivations, but the once-recalcitrant corporate colossus sounds for all world like a True Believer. As then-chief marketing officer Mark (no relation to Jeff) Jarvis puts it, things have changed: "I would say radically. The big revelation for us was that customers did want to interact with us. It's changed our support and our services organization. It changed our product-development organization. It's absolutely changed marketing in the company.... Traditional marketing no longer works. The most important marketing vehicle right now is the consumer. Things have pretty much gone bottom up at Dell."

And Jeff himself feels quite certain this is the real deal: "Note that the company is following suggestions that customers make. At the behest of the geekier geeks among them, Dell is now selling Linux computers and is reducing the bloatware that constipates new machines. Michael Dell acknowledged that the customers' suggestions may not be economically rewarding, but he can't know without trying. Dell even talks about collaborating with his customers. He told me: 'I'm sure there's a lot of things that I can't even imagine, but our customers can imagine. A company this size is not going to be about a couple of people coming up with ideas. It's going to be about millions of people and harnessing the power of those ideas.'"

Friends in Low Places

On November 8, 2007, two months into the project, I got a comment on Comcastmustdie.com: "I work for Comcast. Sorry Bob you are pretty much off the radar now." That might have been true. After an initial burst of activity, comment traffic had decidedly slowed. Perhaps Comcast believed that the novelty had worn off, but there are things Comcast didn't know:

One is that online insurrection isn't a novelty; it is already a fact of business life. Secondly, I was by no means finished. At that point, I was deep into planning for an event that I hoped would hold the corporate feet still closer to the fire of mounting consumer rage. And finally, as I observed 9000 words ago, I'm pretty easily pissed off, but not so easily discouraged. For instance, this is not my first media crusade. Long ago, in the first Clinton term, I tried to take my kids bowling one Saturday only to find every local lane filled up with league play. This rankled me, because I knew there was a perfectly good bowling alley sitting absolutely idle just 20 miles from my house in suburban Washington, DC. It was then that I decided I would take my kids there.

That was sort of a quixotic tilt, too, given that the alley in question was situated in the White House for the exclusive use of the first family and their friends. And I didn't know the Clintons — or, at least, they didn't know me. So I set out calling on friends, acquaintances, acquaintances of acquaintances and anybody I could think of in or around the corridors of power — from my congresswoman to James Carville to the head of the American Bowling Congress to Ralph Nader — to wrangle an invite. This was done for NPR's *All Things Considered*, as a sort of parable of political influence in Washington. In fact, it was essentially the mirror image of Comcastmustdie.com, which is about gaining influence not by being politically connected, but simply by being digitally connected, about having lots and lots and lots of friends in low places.

But there's something you should know. Though the bowling crusade was meant to span the entire summer, the family and I got access to the alley within two weeks. It was fun. And on the subject of the arrogance of power, and its just desserts, I think I wore Nixon's shoes. This got me to thinking about what I'd do if I were in Comcast's shoes. What I decided was that I'd put angry customers back on my radar.

Comcast Must Die: The Podcast

The first thing was the jingle.

I mean, anybody who's anybody has decent theme music. Budweiser. *The Sopranos.* The president. So I advertised on the site for entries for a song in any genre, so long as it conveyed the right message — namely that Comcast must die. But we didn't get one exceptionally brilliant entry. We got two exceptionally brilliant entries, one by a guy named James Cobbins, and another by a guy named Shawn Who. (Both tracks can be streamed or downloaded at Comcastmustdie.com). James' version is full of crashing, thrashing metal chords, synthesizer and drums.

> *Comcast Must Die-ie-ie. So many reasons why-y-y.*
> *They're so aggravating. They just keep me waiting.*
> *They're slay-ing meee, Triple Play-ing meee.*
> *Please why oh why? Comcast must die.*
> *And I can only sigh-igh-igh. Comcast must die!*
> *They just lie-ie-ie. Oh so broke am I.*
> *Comcast must die! Comcast must die! Comcast must die!*
> *COMCAST MUST DIE!*

OK, lilting it isn't, but it is quite a toe-tapper, and you surely can't argue with the sentiment. Meanwhile, Shawn Who and beat producer Antisoc blended techno and hip-hop for an even more percussive approach to the theme:

> *Comcast may claim/*
> *with bombast that they/*
> *can outclass any company you can name./*
> *So who's to blame when y' home all day/*
> *and you sit an' wait, cable guy never came!/*

When they do finally get up in ya house/
see the tech passed out, sleepin' on the couch/
Say the service they got is so "Comcastic"/
but customers say it's "ultra-spastic!"/
Resort to reportin' somethin' drastic/
Comcast has gone way past gettin' its ass kicked!

I'll tell you why/Comcast must die/
It's cuz they lie/ Comcast must die/
We're the thumb in their eye/Comcast must die/
Thanks for droppin' by/ to Comcast must die/

I'll tell you why/Comcast must die/
It's cuz they lie/ so Comcast must die/
We're the thumb in their eye/Comcast must die/
Comin' atcha live/Comcast must die

Let's just say that mere transcription doesn't do these tunes justice. But anyway, the exercise wasn't just for the fun of it. After all, it would have been awfully silly to produce an entire one-hour podcast and not have the proper musical background. Yes, *Comcast Must Die: The Podcast.*

Not a bad idea, eh? Well, perhaps in theory. CMDTP was actually designed to be a live event, streamed online, and thereafter downloadable as a podcast, and probably the first of its kind: a one-hour interview/call-in show entirely dedicated to the proposition that Comcast customers were mad as hell and not going to take it anymore. There were four guests: blogger Jeff Jarvis, Mona Shaw (The Hammer Lady), satirist Harry Shearer and Ralph Nader, each taking on an appropriate slice of the story. Mona spoke about her vandalism episode, Jeff about the web's ability to coalesce anger and support, Harry about his personal frustrations with soulless monopolies and Ralph about his techniques for slaying corporate dragons.

Again, I didn't do *CMDTP* just to be novel. I had a number of specific goals: 1) offering yet another platform to Comcast victims to vent their spleens, 2) acquiring a list of registrants to contact later when their attention or action might be particularly needed, 3) to demonstrate that Comcastmustdie.com is more than a mere hate site, and 4) to grab the attention of bloggers and traditional media, whose buzz would drive more people to the site.

Let's call the event a mixed success. It was lively conversation (you can also stream or download it from Comcastmustdie.com), but the live streaming was a fiasco. Many who logged in could hear only distortion, and many others who tried to phone into the Santa Fe, NM studio — where the production was originating and where a blizzard was playing havoc with the phone service — got only busy signals. Alas, the irony of my failure to make all the technology work was not lost on the trade publication *Multichannel News*, which ran the following web item: "The backers of the 'Comcast Must Die' web site held an inaugural podcast Dec. 11 as a forum for the subscribers to strategize in their fight to receive better customer service, but it was a good thing host Bob Garfield booked some guests. Only two consumers called in during the hour-long podcast." A few gloating Comcasters posted comments on my site along the lines of, "Nyah nyah nyah nyahhh nyah!" Here's one of the nicer ones:

> *I'm a Comcast employee and I just LOVE the fact that you had issues with your call-in system. Now, I'm not saying that there aren't things that my company couldn't improve upon — many of which came from suggestions on this very forum — but maybe now you begin to see the difficulty in running a GIGANTIC network that must meet the needs of hundreds of thousands of customers. Just one caller gets denied access to your show, and now he makes a blog. Was it necessarily your fault? No — but still you are to blame.*

Fine. Point taken — although I wasn't charging anyone for anything, and the entire enterprise was handled not by a multibillion-dollar corporation but by Bart Wilson and me in our spare time. Furthermore, my principal goals were certainly achieved. Thanks in part to the publicity surrounding the event — most especially a feature in *USA Today* — Comcastmustdie.com did attract more traffic, and if I had somehow slipped of the company's radar, I was definitely back on it. For instance, if you Googled "Comcast must die" or similar phrasing, you found an ad from the company touting Comcast Customer Care.

The Five Stages

That was heartwarming. I had spent a couple grand of my own money for the website and podcast; it was nice to know that I was most likely costing Comcast quite a bit more. But I was also by no means finished. Early on, Jeff Jarvis had mentioned something to which at first I'd scarcely paid attention. As I am supposedly an expert in the art and science of television advertising, shouldn't I create a TV spot taking Comcast to task?

It was an interesting suggestion, but also, it seemed to be, a bit impractical. Making 30-second spots is expensive, time-consuming and generally impractical for somebody with two jobs, a half-written book and a consumer jihad to heroically lead. But the idea kept gnawing at me. For 25 years, I've made a living deconstructing ads, often arguing vociferously against gratuitous expense and spectacle. Surely I should be able to cobble something together that would not only serve the website but demonstrate that the secret of doing a good video spot is not to throw money at it. And so, with my own hectoring-pundit voice ringing in my head, notwithstanding my skepticism about Consumer Generated Advertising (see Chapter 8, "Off, Off, Off Madison"), I roughed out an idea.

The premise was to depict a woman on the telephone, in various stages of emotional extremis. In one shot, she is shouting, in another begging, in another praying and so on, dissolving one into another. It was meant to be an Elizabeth Kubler-Ross sort of thing, but instead of the stages of grief, it would be the stages of Comcast-style customer-service aggravation. This I sent to my brother David, who makes funny biographical videos for people celebrating weddings, birthdays, retirements and so on. In two days he refined, shot, edited and mixed "The Five Stages," a dramatization of the anguish attendant to getting your cable TV service fixed. David added a diabolical twist at the end — the actress, Olga Rosin, goes all succubus on us — and briefly blood drips from the Comcastmustdie logo. It's a pretty impressive piece of work (I'd have given it three out of four stars in my AdReview column. YouTube users give it 4 1/2 stars out of five.) Not only did it splendidly capture the experience of trying to deal with Comcast, it captured the attention of the media. Comcastmustdie.com had already been featured in dozens of news stories, including a cover piece in *BusinessWeek* about consumer vigilantes. But within a few days of "The Five Stages" going up, ABC's *Nightline* led its show with my jihad. This triggered hundreds of blog mentions and a flood of traffic to the site.

The total isn't necessarily mind-blowing. Since its March 11, 2008, when we started keeping track, the site has averaged 400 unique visitors a day. What's extraordinary is that thousands of them have left detailed comments recounting their personal Comcast nightmares, many of them truly bloodcurdling tales of horror. What's more extraordinary is that many hundreds of them took my advice and included their Comcast account numbers. And what's most extraordinary of all is that Comcast, as far as I can tell, followed up on every single one of them. Usually within 24 hours. Far from letting Comcastmustdie.com fall of their radar, they incorporated it into their own Listenomics infrastructure, such that it is. Here's a follow-up comment harvested from the site as I write on May 20, 2008:

> *My name is Jonathan and I live in East Petersburg PA. I posted some issues I was having with Comcast on this site and a week later was contacted by their customer service department. I am happy to say that a Tech Supervisor was at my home the very next day after the phone call and fixed my problem completely! Everything has worked as it should since. This site works, everyone! I am proof. Thanx for this site for giving us an outlet to vent and thanx to Comcast for finally listening!*

Yessiree! Thanks Comcast for being so bereft and so incompetent that you are reduced to using a website called Comcast Must Die as your customer-service channel of last resort. Thanks, Comcast, for enduring the abject humiliation of assigning personnel to monitor the primal scream therapy of your worst enemies. And thank you for dealing with your crippling image problems not so much by improving your operations as locating the squeakiest online wheels and coating them with grease. Not just on Comcastmustdie.com, either. In the spring of 2008, Twitter users who had sent their "followers" messages bitching about Comcast were stunned to get return messages from a Comcast customer-outreach manager named Frank Eliason volunteering to intervene. Some people found it vaguely creepy, on eavesdropping grounds, but most were inclined to give Comcast its props. Me, too.

As documented in both *The New York Times* and the *Washington Post* almost exactly a year after the launching of Comcastmustdie.com, wheel-greasing was not the end of the story. In the course of that year, a corner was clearly turned in corporate culture, in some cases precisely mirroring Comcastmustdie.com's explicit demands:

1) Appointing a quality czar, senior VP Rick Germano, to monitor all customer-service issues and build an infrastructure for preventing problems and addressing them. Frank Eliason's team is but one part of that structure.

2) Assigning an in-house team to pore over the internet looking for signs of trouble and discontent.

3) Creating mechanisms for real-time communication between customer-relations employees and repair/install dispatchers.

4) Shifting the incentives for frontline employees from "productivity" to quality; i.e., getting the problem solved versus getting on to the next caller/service order at the expense of the current one.

5) Resolving to host ("within a year," according to Germano) some equivalent of Comcastmustdie.com on its own site, rather than depend on a third party to entertain the criticism, frustration, anger and suggestions of its customers.

"We get it," corporate spokeswoman Jenni Moyer told me. "And not only do we get it, we're not just listening. We're also changing the way we do things. And we're moving from being reactive to proactive." To my surprise and infinite satisfaction, nobody there was claiming that the work was done, or even nearly done. As Germano euphemistically framed the situation in August 2008, "There's a lot of upside for us." Still, the promise to build the online mechanisms for actually hosting the angriest complaints from customers is an astonishing one. Unthinkable, really, a very short time ago. From Germano's perspective, Comcastmustdie.com has been a double-edged sword. In certain respects, he says, "I wish it had never happened." But he acknowledged that it was part of "a bigger wake-up call." I asked him when he'll be ready to cry "uncle."

"Bob," he replied, "I'm crying 'uncle' now."

Shame: the Franchise

And so, my work complete, off I rode heroically into the sunset, bequeathing the website to Bart Wilson in Santa Fe, and looking for other dragons to slay, other windmills to joust, other beleaguered ranchers to rescue from outlaws and corrupt railroad men.

OK, maybe not exactly that. What I specifically did was head for my couch, with a bowl of pretzels and a beer or twelve to catch up on the baseball season. But before resuming my life of idleness and carbohydrates, there was one last errand: the expansion of ComcastMustDie.com — under the new name Customer-Circus.com — to provide online relief for beleaguered ranchers north and south, from sea to shining sea. CustomerCircus offers the same outlet for aggrieved customers of Dish Network, American Airlines and Bank of America as it did for those of Comcast. The same format, the same instructions, the same court of public shame for transgressors. Nor will it be alone. When the vaunted iPhone turned out to be infested with bugs and software limitations, a site called pleasefixtheiphone sprung up and has generated more than

800,000 votes for a variety of fixes. As of this writing, Apple has done 133 of those fixes via software updates.

As I was saying a while back, never pick a fight with someone who buys zeros and ones by the barrel. Which, nowadays, is everyone.

GUESS

"I have the advantage of knowing your habits, my dear Watson," said he. "When your round is a short one you walk, and when it is a long one you use a hansom. As I perceive that your boots, although used, are by no means dirty, I cannot doubt that you are at present busy enough to justify the hansom."

"Excellent!" I cried.

"Elementary," said he. "It is one of those instances where the reasoner can produce an effect which seems remarkable to his neighbour, because the latter has missed the one little point which is the basis of the deduction."

— Arthur Conan Doyle, 1893

HEY, MARK ZUCKERBERG, LISTEN UP. This is about how to monetize Facebook. The billion-dollar answer awaits you just below. Just hang on for about 5000 words, and no skipping ahead, please. That'll just ruin the suspense.

For the moment, though, for all the other faces out there, a little review:

For the past six chapters, I've been railing on and on about a Post-Advertising Age, one in which mass marketing will not depend solely, or even much, on mass media. Pretend, for argument's sake, that I haven't entirely lost my mind. Without assaulting you with the latest network-TV-audience figures or rubbing your nose in the horrifying implosion of the newspaper industry, I ask you to give me the benefit of the doubt. Assume that, in the near future, connections between marketers and consumers will not be principally forged via display advertising, but will be otherwise cultivated online. Assume that technology will offer more and more highly refined means for the marketer to learn about the consumer, and the consumer to enjoy a real benefit in exchange — search and widgets being Exhibits A and B.

Thank you for so stipulating. But if the *lingua franca* of our online future is indeed personal information, where will that come from?

Obviously, a staggering amount will come from the consumers themselves. The *quid pro quo* between the marketer and the audience, for several centuries, has been free or subsidized media in exchange for inundation with ad messages. Madge didn't say "You're soaking in it" for nothing. In the Brave New World, and already in the last remnants of the cowardly old one, the value proposition will be similar, but the barter items very different. A marketer needn't pay for episodes of *Gunsmoke* or *Married with Children* or *24*; it need only provide value — whether in entertainment, information, discount or utility. In exchange, the consumer surrenders data. This is already taking place on an enormous scale. Every online registration you fill out, every cookie you accept on your hard drive, every supermarket purchase you make with your discount card is a something-for-something transaction. This new data economy has obvious privacy implications, but privacy is not an absolute. It is increasingly a commodity — one that celebrities trade for fame, travelers trade for security and most all of us trade for a few pennies here and there, scarcely blinking an eye. We get 50-cents off on a can of New England clam chowder and Safeway finds out exactly how much Preparation H we buy, and exactly how often. And this doesn't even begin to consider what kind of intimate details we post on Facebook. It's a bit eerie when you think about it, but most of us don't think about it. We accept the tradeoff, take the money — or utility — and run.

Volunteered data, priceless as it is, nonetheless takes a marketer only so far. To create a genuine bond, an intimate relationship, requires a thorough understanding of consumer behavior, consumer interests, consumer sentiments, consumer moods, consumer movements and so on — not the sort of information that you can put in a sign-up form, even if anybody were patient or generous or honest or self-aware enough to part with it. This requires what Sherlock Holmes called deduction. Also a bit of extrapolation, inference, intuition, divination, prediction and imputation. Or, put another way:

Guesswork.

Beer and Diapers

It's the stuff of legend, the famous discovery by Wal-Mart that placing beer and diapers near one another in their stores increased the sales of both items. And why? Because men on their way home from work, instructed by their wives to pick up a bundle of Pampers or whatever, also grabbed a six-pack. It is the quintessentially unexpected correlation, almost universally invoked to exemplify the rewards of data mining. Isn't it after all, buried just below the surface, a superficially counter-intuitive connection that makes absolutely perfect sense?

Of course it is. And here's something else it is: not true. The diapers 'n' beer anecdote is marketing apocrypha, a kind of MBA's urban legend, the gerbil-up-the-wazoo of data mining. Usually it is attributed to Wal-Mart, sometimes to 7-Eleven, but the provenance of the story has long since been tracked down. Back in 2002, Professor Daniel J. Power of the University of Northern Iowa traced the correlation to a 1992 data analysis by Teradata Corp for Osco drugs — an analysis of that observed the purchase affinity, between 5 p.m. and 7 p.m., of the two apparently unrelated inventory items. Though Osco did not act on the observation in its store layout or promotional efforts, the striking simplicity of the example just begged for embellishment. "Beer and diapers" has since become not only the gerbil rumor of marketing, it has been enshrined as data mining's Isaac Newton and the falling apple, or Archimedes in the bath. It simply shouts "Eureka."

And that's because, in addition to being false, it is also true.

Never mind that Osco at the time saw no advantage in re-arranging its stores (Huggies in the refrigerator cases?) to accommodate a two-hour-per-day phenomenon. The fact remains that thick veins of unseen correlations lie just below the surface, just begging to be chiseled out and exploited.

"Even though those are seemingly uncorrelated purchases, you can develop pictures of people," says Matt Ackley, vice president of net marketing at eBay, where every ad buy and increasingly every individual user experience is informed by the user's previous behavior on the site. Sometimes, this manifests itself in an obvious way.

"Let's say you'd been on eBay three days ago," Ackley says, "and searched for a particular term. We store that in the user's cookie, so when we see that user out on the web again, and we're serving an ad, on Yahoo Mail for instance, we'll see that cookie. What we then do is pass that information to our banner ad. Now our banner ad is not a banner ad at all. It's a miniature application. And what it does is then goes to eBay and finds items that are like that keyword and pools them into the banner ad."

But beyond ad optimization, there is so much more going on. For instance, eBay algorithms intuit gender from the user's first name and age from the shopping categories chosen.

"We know young people buy iPods and older people buy Longarberger baskets. This is the type of information you can tease out. Well, if you know somebody age, and somebody's gender and what kind of categories they're active in, you can more or less predict what they might be looking for next."

In that way, he says, not only do the online ads change, the eBay experience itself will actually change based on who is logged in. "We know that for certain keywords in certain categories, people tend to be more value conscious. eBay search has a sort order. There are different things you can sort on [lowest cost first, auction deadlines, auction versus fixed price, new versus used]. Well, if someone is value conscious, we will drop them on a page that has lowest cost first. It's very rudimentary, but you can imagine as we go forward."

At a bare minimum, he says, the old saw about "half my advertising is wasted, but I don't know which half" — attributed variously to John Wanamaker, Lord Leverhulme and F.W. Woolworth — will be rendered irrelevant.

"I think we're getting close to solving Wanamaker's conundrum," Ackley says. "We can put that to rest."

Is That Any Way to Behave?

We've stipulated already that display advertising as a marketing tool — inefficient and universally resented as it is — is headed toward near-obsolescence. The key words in that sentence are "headed" and "near." This is a long process, and it would surely behoove marketers to maximize advertising's utility in the interim. They certainly have glommed onto search in a big way. At $8.7 billion in 2007, according to the Internet Advertising Bureau, it represented 41% of all online ad sales. And no wonder. It is geared to intent. If someone is searching for information on the Ford Mustang, duh, they're likely to be in the market for a Mustang. So Ford will serve them an ad on Google, Yahoo, Edmunds. com or whatever. This is what is often called a "no-brainer."

Except that it's really only a half-brainer. Because, strangely enough, purchasing behavior is not necessarily guided by or even always correlated with shopping behavior. Rather often we buy on impulse, which is not nearly so impetuous or capricious as it sounds. It often boils down to being reminded, more or less serendipitously, that a product will fulfill a desire or need. Nobody ever Googled "sodium percarbonate stain removers," but a pitchman hollering "BILLY MAYS HERE" at 4 a.m. stimulated enough latent interest to sell bazillions worth of OxiClean, and lots of other direct-response crap. The problem is, encountering Billy at just the right moment is a matter of chance. And the chances aren't especially good. Even the most finely tuned traditional media buy is based on statistical projections of crude audience data and cruder still assumptions about the proclivities of that presumed audience. In short, it's guesswork, too, but — as John Wanamaker or whomever well understood — a wild, wild crazy-ass guess. No matter how loudly Billy screams, only the narrowest slice of the dazed actual viewership will undergo the epiphany of how handy and affordable a tub of Oxi-Clean might be. SO LISTEN UP! BOB GARFIELD HERE WITH AN INCREDIBLE OFFER YOU WON'T BELIEVE:

"Now we have the ability to automate serendipity," says Dave Morgan, founder of Tacoda, the behavioral-marketing firm sold to AOL in 2007 for a reported $275 million. "Consumers may know things they think they want, but they don't know for sure what they might want. They're not spending all their time hunting for those things."

For instance, flat-panel TVs. In 2006 Tacoda did a project for Panasonic in which they scrutinized the online behavior of millions of internet users. Not a sample of 1200 subjects to project a result against the whole population within a statistical margin of error; this was actual millions. Then they broke down that population's surfing behavior according to 400-some criteria: media choices, last site visited, search terms, etc. They then ranked all of those behaviors according to correlation with flat-screen TV purchase.

In that list, "shopping online for flat-panel TVs" ranked 22 — 18 places below "consumed 'Miami travel' content." Miami travel?

"Not Chicago travel," Morgan says. "Not Europe travel. Not business travel. Don't ask me why. But here's the incredible thing: number one — and significantly above the others — was 'people looking at military content.' It made no sense. Then I talked to a friend of mine who had been an officer in the first Iraq war. I said, 'What's going on?' He said 'That's easy. The kids in the military are huge video gamers. They get big fat signing bonuses and their housing is free. They don't need cars. So they buy big TVs.'"

Morgan followed up because he was curious, and felt the need for this counter-intuitive association to have an explanation. But he needn't have. Why ask why? The whole point is that data mining takes us to a realm beyond obviousness and common sense. The data speak for themselves.

This message was hammered home in research the same year for Budget rental car's weekend-rental promotion. "Shopping for a rental car" was the number four correlation. Number one was "recently read an online obituary." Try to connect the dots if you wish; meantime, go read some online obits and see what ads show up on the page.

"We no longer have to rely on old cultural prophecies as to who is the right consumer for the right message," Morgan says. "It no longer has to be micro-sample based [à la Nielsen or Simmons]. We now have [total-population] data, and that changes everything. With [that] data, you can know essentially everything. You can find out all the things that are non-intuitive, or counter intuitive that are excellent predictors.... There's a lot of power in that."

It is just that power, of course, that horrifies privacy hawks, such as the Electronic Frontier Foundation and the Center for Digital Democracy. The idea of individuals being targeted, to them, is *prima fascie* evidence that individual privacy is violated — or at least put at risk — by behavioral targeting. And there is a certain familiar logic to that. It was precisely Yossarian's logic in *Catch-22*.

> *"They're trying to kill me," Yossarian told him calmly.*
> *"No one's trying to kill you," Clevinger cried.*
> *"Then why are they shooting at me?" Yossarian asked.*
> *"They're shooting at everyone," Clevinger answered. "They're trying to kill everyone."*
> *"And what difference does that make?"*

As far as Captain Yossarian was concerned, as long as enemy gunners were aiming antiaircraft flak at his plane, they were trying to murder him — which on one level scans. But Yossarian was regarded as crazy for failing to understand that no artillerymen had it in for him personally, Capt. Yossarian, the individual. They merely wanted to shoot down the Allied bomber, and the plane's navigator with it, whatever his name might be. It's the same with behavioral marketing. The difference is, those whose IP addresses are targeted online aren't shot out of the sky. They're shot some less irrelevant ad messages.

Flicked Off

Here we are in the Silicon Valley — Los Gatos, to be exact — at an intersection of Winchester Circle and California 85 that is nobody's idea of FutureWorld. No shiny glass and steel edifice, no robots, no ports for hovercraft. It's a nondescript, latte-colored Spanish stucco soaring a majestic three stories high, across the street from the freeway on-ramp. The chain-link fence on the perimeter of the office park is decidedly low-tech, but it needs to be there; otherwise, pedestrians would be put at risk every Friday at 2:30 p.m. when, on the parallel Southern Pacific Railroad tracks, the coal train goes chugging by.

What valley is this again? Ohio?

Enter the corporate headquarters here and there's still no way to get a firm fix. It's the usual warren of cubicles spiraling around floor after open floor. As you wend around the maze of cubes, the only sounds you hear are the muted clicks of industrious fingers on unseen keyboards. The eerie quiet is easily explained: most of the company's actual operations don't take place here; they're spread among 55 warehouses (not virtual warehouses, just warehouses) around the country. So, yeah, everything about the place screams "typical back office." You could be in the sales hub of a cooling-tower manufacturer in Akron or a claims-processing center in Hartford or some far-flung auditing gulag of the IRS. If it weren't for the red popcorn wagon in the ground-floor reception area and the little conference rooms dotting the place with names like *Batman*, *Jaws*, and *Apocalypse Now*, the environment would tell you nothing at all.

Unless you close your eyes and feel the rhythm.

I mean, the algorithm. This is Netflix, the company that built a substantial business by delivering movie DVDs overnight at a flat fee, and built a gargantuan business by recommending to customers — via the miracle of collaborative-filtering software — what movies they'd like to see. Be not misled by the stucco and ugly teal industrial carpeting. This is a technology company to its core.

"Harnessing the power of the community to generate better results for the individual."

That's Reed Hastings, founder and CEO of Netflix, who was describing not only his company's methods but also the essence of collaborative filtering, which is one of the ABCs of predictive marketing. B is behavioral, which is the ability to track your path online. C is contextual, which pays attention to key words, and A is associative, which can divine your tastes and interests based on patterns established by people like you. When Hastings and his partner Marc Randolph started the company in 1998, it was an *à la carte* DVD-by-mail service at $4 per rental, free shipping and no late fees. This was reasonably successful, since many Americans were paying, like, $34,000 per month to Blockbuster for the VHS of *Pretty Woman* lodged under the minivan seat. The business picked up substantially a year and a half later, when Netflix introduced the subscription model: (most popularly) three titles at a time, for as long as you wish, for $16.99 per month. But it truly leapt forward when, in 2000, they incorporated their proprietary Cinematch recommendation engine, which scans a user's rental history, her ratings of those films and her ratings on an ongoing Netflix survey to make suggestions on movies she'd likely enjoy. Anybody who has ever browsed the video store for an hour and walked out empty-handed — or, dispiritedly, with a DVD of *Mr. and Mrs. Smith* — will immediately understand the benefit. My personal Netflix recommendations fall into two categories: movies I've seen and loved, and those that I haven't seen ... but sure will make it my business to, because the algorithm has clearly anticipated my tastes.

If Netflix can figure out I admire *Manhattan, Strictly Ballroom, Happiness, The Girl in the Café* and *What Iva Recorded on October 21, 2003*, how badly can *Rabbit Proof Fence* and *Fitzcarraldo* disappoint me? The consequence is a great boon to me: easier selection process, fewer duds. It's an obvious boon to Netflix, which had 239,000 subscribers when Cinematch was launched versus 8.4 million today. And it is a veritable godsend to the movie industry. Not to the Hollywood-studio part of the industry. *Spiderman 17*, or whatever, will do just fine on its own. Netflix's impact is on cinema's everything else, the so-called Long Tail of moviemaking.

The Long Tail is the coinage of *Wired* editor Chris Anderson, whose seminal 2004 magazine article on the subject yielded an ongoing blog about it, which in turn yielded a best-selling 2006 book about it — the "it" being how digital technology has ended the near-monopoly on distribution enjoyed since the Industrial Revolution by mainstream blockbusters at the expense of niche goods and services. The fat head of the Long Tail is *Spiderman*. Way, way, wayyyy down in the skinny middle is *Fitzcarraldo*. But now I can rent them both in one click.

This is a testament, of course, to the internet's vaunted democratization of everything. At the point of online rental, Spiderman's $100 million production budget confers no particular advantage. Maybe at (the aptly named) Blockbuster, which fills its shelves with 4000 copies each of eight new releases and no copies at all of *Rabbit Proof Fence*, the fat head still rules. But Anderson's thesis is that in an online universe the fat head inevitably will get thinner, and the Long Tail will plump right up. And right in the middle of the transformation is collaborative filtering. Because, in terms of connecting consumers with what they actually want, it is simply a better mousetrap.

Till now, when our choices were (as a practical matter) limited to what The Powers That Be said our choices would be, filtering wasn't collaborative, it was unilateral — or, as Anderson describes it, "prefiltering." This means experts — record labels, movie studios, publishing houses and so on — bringing experience, instinct and time-tested judgment to bear on the process of selecting content for the public. Whoever signed the deals for *Titanic*, the Beatles, *Cats* and *The Da Vinci Code* were obviously shrewd judges of quality, or at least mass tastes. But here's the thing: experience, expertise and judgment aren't the same thing as clairvoyance, even when the geniuses also command *de facto* monopolies on production and distribution. More or less the same class of visionaries also invested in Broadway's legendarily unwatchable *Moose Murders*, O.J. Simpson's aborted literary meditation *If I Did It*, backup dancer Kevin Federline as a solo recording artist and, in the same 1997 as *Titanic*, Kevin Costner's calamitous vanity pic *The Postman*. Hence the famous Hollywood axiom is "Nobody knows anything," which aims to explain, for example, how Warner Bros. could have spent $120 million on production and $60 million on marketing for *Speed Racer*.

Furthermore, the financial pressure to be right forced the Judgment Gods to err on the side of (presumed) mass appeal, which is also called the lowest common denominator. That phenomenon does explain why *Independence Day*, by many orders of magnitude the worst movie ever made, was green-lighted by 20th Century Fox. It also explains why it was a big hit worldwide. By definition, exactly half of the people on earth are dumber than average, but they still get to buy movie tickets — a structural reality that has historically worked against those of us with a taste for Kurosawa and documentaries, or for that matter, for those with a taste for anime or Christian romance. In other words, economic necessity created a marketplace that not only limited distribution of niche products, but also suppressed their production altogether.

Compare pre-filtering, then, to post-filtering — collaborative filtering — which, with the ultimate benefit of hindsight (it operates only in hindsight), knows everything. This is especially useful in a digital, Long Tail universe of seemingly infinite choices. Like my friends, former Soviet refugees, who walked into their first American supermarket and burst into tears, we are easily overwhelmed by the astonishing array of items on the internet's virtual shelves. This phenomenon is often described as "information overload," but Clay Shirky, author of *Here Comes Everybody* and professor of new media at New York University says that's not quite right. We suffer, he says, "not from information overload but filtering failure. The minute people are exposed to reality they freak out. What collaborative filtering does is replace categorizations with preference."

In the example of Netflix, like bricks-and-mortar video stores, the Cinematch recommendations are divided by genre: drama, comedy, foreign and so on. But within those genres, they aren't sorted by alphabetical order, like the titles at Blockbuster; they're broken down by what your stated preferences, your viewing history and the preferences of people with viewing histories much like yours. And by no means is that technique limited to movie rentals. It is the mathematical basis for Amazon.com's book recommendations, for Match.com's e-yenta service and every "people who bought lox also bought bagels" message you've ever seen on an e-commerce site.

"If you give us all of your content," says Paul Martino, founder of the data-mining provider Aggregate Knowledge, "I'll put it in front of the right person at the right time." It's an interesting boast, because he's not speaking merely of Amazon.com recommending books based on the book-purchasing behavior of you and people with tastes like he. He's talking about recommending barbecue grills and newspaper stories and hair-care products based on the book-purchasing behavior of you and people like you. "It's not just books for books," he says, "It's anything, matching with anything else."

Mark Zuckerberg, hold that thought.

As the Late Billy Mays Would've Hollered ...

... but wait! There's more.

In some applications, collaborative filtering is useful, but insufficient. Think about video sites. Considering that videos usually lack metadata — underlying searchable text — beyond a couple of keyword "tags" volunteered by the uploader, it would seem that "people who watched that also watched this" would be the core of any recommendation system. But it ain't necessarily so. Since most sites prominently display thumbnails of their most popular videos, and since that practice results in a further snowballing of popularity, a pure collaborative filtering approach would likely wind up recommending the same handful of videos again and again. To most publishers, who want users to stay on the site longer and longer so they can be served more and more ads, there is little benefit to a dog chasing its own tail. So how to intuit what, beyond the big hits, a given user would be most interested in seeing?

Visit Israel, a journey of discovery.

Tucked away in one the Tel Aviv R&D Center's many unsightly office-building monstrosities is Taboola, a 2006 startup that—like many of its neighbors—seeks to exploit data-mining techniques of the country's legendary security apparatus for commercial applications. But instead of trying to divine the intentions of Iran or Hamas, it is devoted to doing the same for someone who has clicked on, for instance, 5min.com to watch how-to videos for do-it-yourselfers. Founder Adam Singolda, who commanded a data-mining team for the Israeli Defense Forces and National Security Agency, says it's all about "pattern recognition ... the hidden factors that lead you to perform your actions, that you didn't know yourself."

Some of this does involve collaborative filtering; people who watch circular-saw demos may well be recommended videos favored by other people who watch circular-saw demos. Also considered are textual cues, à la search, matching the most-likely sparse metadata in one video with the most-likely sparse metadata from another. That process also employs so-called "data enrichment," in which certain keywords trigger associated keywords from the Taboola lexicon. For instance, though an iPod video may be tagged merely iPod, or not tagged at all, clicking on it may well get you a recommendation for a video—or may get you served an ad—about MP3 players, or about Apple. The most critical function, though, is to observe the user's behavior on the site. "If you close a video quickly, "Singolda says, for example, "if you skipped past the first three recommended, if you hovered over a thumbnail before choosing it, if you hovered over one without choosing it, if you commented, if you commented twice." All of these behavioral clues are crunched to constantly refine the recommendations.

At 5min.com, one of Taboola's guinea pigs, the results have been striking: 30% more video views, 41% more views of entire videos and a 50% increase in the average user session time on the site. Moreover, the same predictors of relevance and intent are applied to ads served to the site. For competitive reasons and, it says, scientific caution, Taboola refuses to disclose actual click-through rates—apart from asserting, based on very preliminary results, they have increased across the board.

All of which is to say, the utility of any of these technologies is impressive. The potential in combining them is simply staggering.

Let's consider another Israeli startup, this one called My6Sense. Situated not far from Trendum's word-of-mouth mavens in the Herzliya Pituach tech corridor, My6Sense has created a message-ranking engine, facilitating "user discovery" of content based on associative, behavioral and contextual connections, but also on a fourth dimension: location. The goal is to consolidate all the various information streams coming into your mobile phone (RSS feeds, news headlines, emails, text messages, Facebook activities, Twitter tweets and judiciously served ads) and list them for you in order of priority. Not your sense of priority, but the algorithm's sense of your priority.

"We don't believe that people actually know their own preferences," says co-founder and CEO Avinoam Rubinstain.

He cites a classic psychological experiment, the "hat test." Asked whether they prefer a green hat to a white hat, the majority will assert they prefer the green hat. But given an opportunity to actually choose a favorite, the majority will select a white one. In the same way, users asked to prioritize their various mobile-phone feeds may not know their own minds, or at least their own impulses — especially since the relative value of some information changes with location. An English businessman might, for instance, be keen on being alerted "If there's a traffic jam in London. But he doesn't care if he's in Israel."

I asked Rubinstain if his algorithm synthesizes urgency. He shook his head, and reminded me that various mobile alert systems failed because urgency is a subjective matter and pretty much the ultimate moving target. Then he smiled: "We synthesize relevance."

Impressive, no? That's why these technologies are transforming marketing. Or, anyway, some of them are. Search is a monster getting more monstrous with every fiscal quarter. Behavioral targeting has become a basic tool of the trade. Collaborative filtering, though, for some reason lags. Yes, Amazon.com and Netflix are high-profile applications, and online-retail recommendations are more or less ubiquitous, but collaborative filtering as a technology is, for the moment, running in place — especially in the social-networking arena where it would seem to hold limitless possibilities. This even though social networking is itself a direct outgrowth of collaborative filtering.

In 1992, scientists at MIT wished to a) advance the art of programming, b) help the chief researcher with her play list. "I was interested in getting recommendations for music I might like and my tastes are very eclectic," says Professor Pattie Maes of MIT's Media Lab. "It was really simple stuff, to make my life a little easier."

"We've built many variants of the algorithm," she says. "In the commercial system we built, called Firefly, we actually made it possible to send emails to people who had similar tastes. And there were even marriages that came out of it. We weren't trying to make matchmaking service. We just thought it would be a fun component of the website."

Yeah, and Jed Clampett thought he was hunting possum when he struck oil. As NYU's Clay Shirky explains it, the researchers had the concept backwards: "The classic CF paradigm," he says, "is to use people you know to find things you like. But it turns out that people are far more interested in using things you know to find people you like." Facebookers, for instance, make and cultivate connections based on common affiliations — chiefly professional and educational — and on mutual affinities for books, music, TV, social causes and so on.

"All We Need is a Model?" Not Anymore

Which brings us, finally, back to the 24-year-old wunderkind Mark Zuckerberg. Dude, blessed as you are with the megaphenomenon called Facebook, why are you just another popular utility in search of a business model? Could it be that you're fixated on the notion that your revenue must come from typical advertising? Haven't we agreed that advertising is problematic, because users are suspicious of it, because they are resentful of it, and because they employ every means to avoid it? Yes, we have. Yet the same people 1) love goods and services, 2) crave information, 3) are so fabulously self-involved that they display every last detail about themselves, their tastes, their preferences, their favorites, their hobbies, their embarrassing drunken photos, their damn near everything right on your site.

So why in the world do you not have a big honking box on the bottom of every Facebook page titled "What You'll Like" or "YouStuff" or "The Mirror" with a category by category selection of books, music, films, videos, news articles, websites, tennis gear, shoes, power tools, specialty foods, flea and tick protection ... you name it?

Yeah, to the online denizen, increasingly all advertising is spam — but this wouldn't be advertising. It would be a set of objective recommendations informed by all of the associative, behavioral and contextual technologies mentioned above. In other words: content. "The Mirror" would be an application, a house-brand uberwidget. And it would be valued by users in approximately the way the *Playboy* Advisor was valued by generations of striving young men — but exponetially more, because these recommendations aren't crude guesses based on some broad demographic data, advertiser pay-for-play strongarming and the editors' tastes. They're extraordinarily educated guesses based on the most granular personal data ever gathered and analyzed. And they would quickly prove to be indispensable, because from the very first trial, users will think, "Whoa! I do like that. I do want that. Cool!"

It will be *Rabbit Proof Fence*, in other words, times every product category in the world.

Naturally, you wouldn't charge the users. Nor would you charge any manufacturer or retailer or other provider for the listing. But you certainly would charge them for the hyperlink.

The links, Mark. The LINKS.

Good luck, young fellow. But guess what? No need to thank me. It's no big revelation. Really, it's as simple as ABC — or, put another way: Elementary, my dear Zuckerberg, elementary.

SOMETIMES YOU JUST HAVE TO LEGO

A MAN RIDING A GOAT.

A dachshund. A fire truck. A quacking duck. In 1932, when Ole Kirk Christiansen began his toy company, his product line included no electronic robots equipped with sound and motion sensors. There weren't even any lugged bricks; plastic was still futuristic technology. What there was was a worldwide depression and an unexpected marketplace for simple wooden toys, which he, as an out-of-work carpenter, had been reduced to building, along with stepladders and ironing boards, to put ludefisk on the table for his young family. Luckily, apart from economic hardship, the other easy-to-find thing in Billund, Denmark was wood. The woods were full of it.

In addition to raw material and good fortune, there was yet one further blessing — utterly unbeknownst to himself — that Ole Christiansen possessed lo those 75 years ago: the capacity for foreshadowing. By 1934, with Christiansen's shop now boasting 15 employees, he decided to name the company. It was an early, and strangely prescient, exercise in crowdsourcing. He staged a contest, offering a bottle of wine to the winning entry. Surrendering to the will and wisdom of the stakeholders! Listenomics!

Sort of. Turns out, "prescient" is not the same as "sainted." Christiansen, the chief judge, won his own contest. He created a contraction of the Danish words Leg ("play") and Godt ("well") and dubbed the firm Lego. History does not record what became of the wine, but like a billion or so plastic bricks to follow (together or lodged in the sole of Daddy's foot), the name stuck.

The business grew well enough. The quacking-duck pull toy became a childhood icon, and Lego prospered until a catastrophic fire burned almost the entire enterprise to ash — a development that put that old woodworker in a less all-wood-inventory frame of mind. In 1949, Christiansen bought a patent from Kiddicraft, a British firm, for cellulose acetate bricks molded with protruding round studs. That the few shapes were interlocking and universal appealed to the Dane; ever prescient, he envisioned infinite possibilities — for model-building and for the company itself. Alas, he evidently could see the future in a way the trade could not. Toy stores and consumers themselves were slow to adopt plastic, and Lego struggled against their indifference. A breakthrough came in 1958, when Lego advanced Kiddicraft's technology by molding circular "tubes" into the underside of each brick, vastly increasing their ability to grip the topside studs — what Lego calls "clutch power." But the advance took five years and a new plastic material — Acrylonitrile Butadiene Styrene — to fully incorporate. Only then, in 1963, did the Lego system explode worldwide to become the most ubiquitous toy of the 20th or any previous century. (If you divided up all the Lego bricks among everyone in the world, it would amount to 62 apiece.)

The problem, however, was the 21st century. Building blocks are fun, but action-wise they have a hard time competing with, say, Wii. Thus, as the Millennium approached, did Lego contrive to embrace technology with a series of kits called Mindstorms: a combination of Lego blocks, motors, sensors and other components for building programmable tabletop robots. The target: boys 12 and up.

Oops.

Mindstorms launched in August 1998. "A few days after the launch," recalls Steven Canvin, business development manager, "the product was taken apart, an inventory of all the hardware components compiled and the [software] code was revealed."

Boys much upper than 12 had immediately reverse engineered the kit and were sending all the specs hither and yon across the internet. "The lawyers freaked out," Canvin says, "because that was an infringement of our rights. They were afraid we were being trespassed upon." The lawyers of course, weren't wrong. This was not meant to be an open-source line of products. The information suddenly in wide circulation was clearly proprietary. But wait. Winging its way across the internet? Lego? Was it possible that Ole Christiansen's little workshop was creating buzz? Yes, it was. And there is value, incalculable value, to buzz.

"It migrated into the adult community," Canvin says. "We heard so many times, 'This re-created Lego for me.'"

Freaks and Geeks Speak

Management in Billund responded to this development essentially by doing nothing. It didn't sue anyone for intellectual-property infringement, and it didn't change the Mindstorms marketing plan, such that it was. It just left the product on the shelves, where it did just fine, if not necessarily with the target audience they'd had in mind. But six years later, with the original version getting long of technological tooth, the company went online to take a closer look at the community that had clustered around Mindstorms. Lego discovered that the fans had not only embraced the robots but customized them, using various parts and sensors of their own contrivance to give the toys capabilities beyond the off-the-shelf design.

"They know the product better than we do," says Canvin and hence the Big Idea: to recruit a handful of fans for collaboration in the redesign. Using Google to track down their email addresses, Lego got in touch with four hardcore enthusiasts, offering them the chance to participate in the design of the new Mindstorms kit provided they sign a non-disclosure agreement and travel to Billund at their own expense.

"Within an hour, they had all signed the NDA," Canvin recalls. This was the Mindstorms Users Panel, eventually expanded to 13 Lego freaks and geeks, who are affectionately referred to in Billund as Legoholics. How addicted are they to finding novel ways to snap stuff together? Very. "Some of them scare me," Canvin says. Not so much, however, to be denied free rein around Lego's super-double-secret design inner sanctum. A cluttered warren of work tables and overflowing shelving in the heart of the low-slung corporate campus, the R&D center is off-limits to most employees and all outsiders. "We actually brought them inside," Canvin says. "Nobody gets in and out without permission. They got to roam around freely."

Operational security is what guards the keys to the kingdom. Putting it at risk obviously required an unprecedented concession from top management, who are justifiably paranoiac about the surprisingly robust plastic-building-block competition. Sensing that no such concession would be granted, the Mindstorms team handled the situation with the delicacy of middle managers since time immemorial:

They didn't bother to ask.

"It's better to ask for forgiveness than permission," Canvin sagely observes. And so began the process. Fourteen months later, prototypes for the second generation of Mindstorms were developed. Listenomics did not end there, however. On the contrary. At Lego, Listenomics was just beginning.

"We were spearheading a new way of thinking," Canvin says. "It dawned on the company that the community around Lego is so big we really hadn't been taking advantage of them efficiently." For instance, instead of its ordinary practice of hiring a firm to do beta testing of the prototypes, Canvin's group went to online discussion groups and announced that it would create an open web forum of 100 users. They got 10,000 applications. And the lucky 100 paid $150 apiece for their kits. "They're crazy. They're really, really into it," Canvin says, with more awe than condescension. "I don't know how adults can spend so much time doing that, but apparently they can."

The result? A relaunch of Mindstorms to an engaged, enthused, energized audience champing at the bit for the new designs long before they were even in production. The fans consulted along the way were not only co-designers and product testers; they were evangelists, e-fanning far and wide to spread the Good News for Model Man. Needless to add, Mindstorms 2006 was a huge hit.

"It's a fraction of [total corporate] sales," Canvin says, "but in terms of profitability it's the deepest-selling product Lego has ever had. And it created a volume of PR never seen before in the history of the company."

Yes, this was your quintessential PR bonanza — except that, unlike most other quintessential PR bonanzas, it has never stopped. On the contrary, the community that coalesced around the second-generation Mindstorms has stayed coalesced, and grown. To get just a small notion of the intensity of the cult, you might want to check out thenxtstep.blogspot.com, one of at least a dozen hardcore blogs trafficking in the latest on Lego robotics. Just for instance, this paragraph is being written on July 2, 2007. At 2:37 this morning, contributor Fay Rhodes posted thusly:

"I realize that sound files eat up memory in the NXT Brick, but if anyone out there is creating custom sound effects for their robots, I'd like to hear about how you are doing it. What software do you use? What equipment is required? Is there a way to minimize the size of the files? How do you download it to your Brick? Any other advice?"

Within 6 ½ hours, she had received 13 replies. One of them, from Christopher Smith, was this:

I'm making a bunch of custom sounds and I'm finding new ways to squeak any space savings possible. I'm also looking for the best WAV to RSO compactor that will not reduce the quality too much. Mainly because the BenderBot (featured in theNXTstep Idea Book) will use sound files and I hope to have a bunch of good ones by the time the book is released. I cover some of this topic in the book also. I using a great guitar signal effects processor and I'm recording some very cool sounds.

Here's some tips:

Once you have created a RSO file you need to copy it to the proper directory to use it with NXT-G. The directory is: C:\Program Files\ LEGO Software\LEGO MINDSTORMS NXT\engine\Sounds

Once the file is there you can choose it in your sound blocks.

Most music programs will be complex or at least require large sound file sizes. The NXT brick contains a number of programs and files preinstalled which take up about half of the available memory space. These files can be removed and reinstalled with no consequence. Reinstalling the NXT Firmware will completely restore the onboard software.

I recommend maximizing onboard memory space by:

- *Downloading the newest firmware using the NXT-G software.*
- *Deleting unused programs.*
- *Deleting unused sound and graphic files.*
- *Deleting preinstalled files and programs. You should try to remove all of the Demo programs and unneeded sound files (NXT pwr on sound, click, etc.)*
- *Create subroutines to reduce the number of programming blocks needed.*
- *Create programs with Mini Blocks when possible.*

For detailed information about deleting files from your NXT brick as well as how to manage NXT memory and firmware downloads please read the LEGO MindStorms NXT Users Guide or by activating the Help functions within the NXT-G Software.

That there, ladies and gentlemen, is devotion. For a consumer product, mind you. For a freakin' toy! Not only should every company in the world be envious, I myself am envious. Once, in circumstances that left me no other choice, I begged my teenage daughter to drive me to the airport for an international flight. She wasn't quite as thorough and accommodating as Lego fan Christopher Smith. All she said was "No."

Institutionalizing Discovery

Once it dawned on Lego that its customers were happily participating in design, testing, marketing and customer service, it did not take long before it was incorporating Listenomics far beyond Mindstorms.

"We try to use the consumer at the very earliest stage of the first idea," says marketing manager Martin Lassen. "We have this assumption that this increases our chance of hitting the bullseye. If we do something wrong, they spank us immediately."

So now, instead of standing user panels, it hosts three Boost Weeks per year, in which Lego pilgrims from around the world make their Billund hajj. The results have been predictably agreeable for all involved. Consider, for instance, the company's Creator line, complex model kits that can be reassembled into two other forms. For instance, the stegosaurus kit can also create a pterodactyl and a tyrannosaurus.

"We had our designers working on a Creator model, a transport ferry," Lassen recounts. "It was not chosen. It was just sitting on a shelf. But at a Boost Week, five fans saw it and said, 'Wow. It's a very cool model.' I was thinking, 'Yeah, but not for little kids.' But they were very enthused and we decided to give it a try. And at regular product testing it did well with Americans, Britons and Germans." So it was in the next line. So was the Café Corner House, which is as it sounds: a typically European café on a city corner. One of the designers was Jamie Berard, a 34-year-old American who was hired fulltime in Billund after impressing during a 2005 Boost Week. As an employee, he in turn embraced an idea tossed around at a 2006 Boost Week.

"One of the things that kept coming up," he recalls, "was buildings. We have fire stations and police stations, but nobody seems to live anyplace in Lego."

Hence his idea: to have uniform buildings connectible in town layouts, expanding vertically and horizontally. Management was at first very skeptical. Residential buildings had always been deemed too dollhousey for the male target audience, but Berard and others persisted. "I said, 'It isn't a dollhouse. It's an expandable modular building system.'"

To sit and chat with this guy is to marvel at man's capacity to be fascinated by things that, objectively speaking, don't much matter. Strangely, the experience altogether lacks pathos. On the contrary, it's weirdly inspiring.

"I'm a lifelong Lego fan," Berard says. "A lot of adults re-discovered Lego. I've had the good fortune of never leaving Lego."

The guy is slightly built enough and Lego-obsessed enough to invite assumptions about his, shall we say, level of sociability. But any suspicions of crippling nerdiness are quickly disabused. He is a preternaturally affable and outgoing young man, with a ready handshake and easy smile. He also has one of the world's more eclectic resumes. Since graduating Merrimack College in 1999, he's been a surveyor, computer serviceman, carpenter, video editor, deli clerk and monorail pilot at Disneyland — all, he says, to support his Lego jones. Seventy percent of his possessions on earth are Lego. And God knows what percentage of his waking hours. Berard's magnum opus: the 96' x 22' Millyard Project at Manchester, NH's SEE Science Center. He was a foreman on the project that used 3 million Lego bricks to recreate the textile mills of Manchester, circa 1900.

It is not lost on him that even Lego management cast a suspicious eye on the Boost Week boosters — that the Legoholics might be, let's say, a few colorful Acrylonitrile Butadiene Styrene bricks short of a load.

"They had the impression that fans are a little excessive and a little weird," he says. "They didn't understand that there is an emotional connection to the product."

Please note the past tense.

Tasty Murder

Is it easier to tap the emotional connections between a child (or former child) and a toy than between, say, an environmentalist and a locomotive? Maybe, but do not discount the amount of engagement users — or critics or bystanders — have with the most innocuous of goods and services. General Electric, in building a corporate strategy for its more-energy-efficient engines and turbines, did indeed cultivate environment bloggers to get them aboard the initiative. Delta Airlines and Starwood Hotels have created online communities of travelers, who weigh in on everything from the size of airplane blankets to the pricing structure of phone calls. HP does the same with its digital photography business, Frito-Lay with its snack foods, GlaxoSmithKline with a panel of women helping to strategize (and, of course, evangelize) introduction of a new weight-loss pill. Not to mention Starbucks, which didn't offer soy milk until the blogosphere demanded it.

There are many examples just like them. Let us not forget, though, the most satisfying of all: Dell Computers. You read earlier how Dell was embarrassed and seriously damaged when one blogger went e-public with his personal consumer nightmare. After at first reacting like precisely the arrogant, tone-deaf behemoth BuzzMachine.com's Jeff Jarvis described, Dell was eventually chastened by the experience. Since then, it has embraced Listenomics with the zeal of a convert. In February 2007, it launched IdeaStorm (Hmm. "MindStorms?" "IdeaStorm?" Whatever.) with the express mission of soliciting customers comments, criticism and product suggestions. In the first four months, it received 24,000 comments and 5,500 suggestions — 21 of which triggered in-house product initiatives. Among them, the pre-installation on all Dell computers of the Linux-based Ubuntu operating system. But IdeaStorm isn't just an online focus group or virtual suggestion box; it's a community, in which users can vote "yea" or "nay" on others' ideas. As Dell communications VP Bob Pearson told my *Ad Age* colleague Matt Creamer, "With the average focus group, you go in for an hour or two, give them some sandwiches and leave. We may be listening to conversation going on over two months."

Remember how the Lego middle managers gave inner-sanctum access to the Jamie Berards of the world? IdeaStorm was devised, Pearson told *Ad Age*, "to make sure the customer is walking the hallways at Dell."

Walking the hallways and shopping the aisles — what a sweet thought, a Stevie Wonder song of peaceful co-existence giving way to actual intimacy. One of my favorite examples is CafePress.com, which is an open-source, online emporium of printed matter — greeting cards, posters, books and especially t-shirts — that doubles as an entertainment site (some of the t-shirts are hilarious: "Meat is Murder. Tasty Murder") and triples as an online community of writers and designers. CafePress is in some ways like eBay. Ordinary folks can use it as a channel of promotion and distribution for their products, and nobody needs to persuade any novelties-store buyer or casual-separates buyer or shirt-maker that an idea is worthy; you post your design and wait to see what happens. But CafePress is also like Flickr, the photo-sharing site. When your design goes up on the site, it stays there, open to inspection — and reproduction — by anybody. The designer gets a cut of any sale, but to do so he surrenders control of his intellectual property, just as Fred Durham surrenders control over the merchandise he sells. Filtering out only pornography and hate speech, he displays all incoming designs on consignment and prints on demand, letting the community decide what is worthy. This has yielded some surprising hits.

"There was a college intramural basketball team," Durham told me. "Happened to be they were all Native Americans, in Colorado, and they just created this kind of a dorm pickup basketball team called The Fighting Whities. They were making fun of the, you know, kind of "fighting Indian" sort of slogans. And they had this sort of 1950s white guy clipart on the t-shirt, and it turned into a massive media frenzy. Everyone had respect for this thing that felt like it should be controversial but was just funny and amazing and touching and, you know, patriotic all at the same time. I don't know how you could mix more together than with The Fighting Whities."

The Fighting Wikis

Such examples of mining the Collective Ingenuity. For instance: the cover of this book. The design was one of 128 submissions by more than 100 artists submitting ideas for the job on a website called crowd-SPRING.com. Ordinarily, a publisher will throw a manuscript at an in-house designer, or a single freelancer, and live with the results. My project, by contrast, was available to the entire crowdSPRING community of 16,000-some artists and designers, experienced and inexperienced, amateur and professional. Those who read my brief had the benefit not only of my direction, but the efforts of their colleagues, and my feedback to everyone. (At TheChaosScenario.net, feel free to check out some of the submitted ideas, or for the whole portfolio, at www.crowdspring.com/projects/graphic_design/illustration/the_chaos_scenario). So good were the entries and so difficult was the final choice, that I selected three of them — leaving the final decision to visitors of TheChaosSenario.net. The cover on this book was sourced to one crowd and selected by another. This was possible because, not only did the free-for-all vastly expand my array of options, the competition among many suppliers by its nature drives down price. I offered $500 for a book cover. Even with my decision to give that sum to each of three artists, my bill was a fraction of what I'd have paid to a bricks-and-mortar design shop weighed down with overhead and payroll and stylish, uncomfortable Italian furniture. As you'll see in the next chapter, the democratizing of supply is a double-edged sword. The economics of commissioning work from outside of the walled gardens of established institutions has the potential to undermine the business model — i.e., destroy — entire industries. Such, though, is the toll of revolution. Perhaps in such times it is natural, if you have cultivated and depended upon the bounty of that garden, to reinforce the walls. But, of course, that is no protection against the encroaching violence. Survival indeed depends on venturing outside the perimeter, as a first step toward tearing down the walls altogether.

In the last chapter, you read about Cinematch, the Netflix movie-suggestion engine. As I illustrated, Cinematch is very, very good. But it isn't good enough to suit Netflix, which understands that credible recommendations are what keep subscribers coming back for more. Naturally, Netflix has its own mathematicians working the problem. But how much better and faster the results might be if every mathematician in the world were on the same hunt. So Netflix has offered a prize of $1 million to anyone who can improve the reliability of the company's collaborative-filtering algorithm by 10%.

(Mind you, I spoke personally to Netflix's Reed Hastings about that, but just to double-check some basic facts — before writing the last paragraph — I consulted Wikipedia, perhaps the most crowdsourced anything anywhere. According to Wikipedia, "Wikipedia's 12 million articles ... have been written collaboratively by volunteers around the world, and almost all of its articles can be edited by anyone who can access the Wikipedia website. Launched in January 2001 by Jimmy Wales and Larry Sanger, it is currently the most popular general reference work on the internet." It's also the 7th most-visited website in the world. *Encyclopedia Britannica* had a lovely garden, long since trampled by an extremely well-informed mob. Even Microsoft's online Encarta was no competition. In March 2009, in grim acknowledgment of Wikipedia's 97% online market share, Microsoft pulled the plug.)

Striking Oil

Not that crowdsourcing is by any means confined to facts we actually, cumulatively possess. It also turns out to be dandy for figuring out things we don't know. A big R&D division is all well and good, but if you require some problem solving, doesn't it make sense to expand the exercise beyond your own corporate labs and research team to, say, all the world's minds, whether in your employ, papered by doctorates, certified in the particular field of inquiry or not? Of course it does, which is why no less an institution than the British government embraced the strategy — in the year 1714.

The stickler in question was longitude. Thanks to astrolabes, the greatest maritime power in the 18th-century world had no difficulty ascertaining latitude at sea. The north-south axis, however, was a much stickier wicket, and so parliament offered 20,000 pounds — the equivalent of many millions of today's dollars — to whomsoever could produce a workable solution for mariners. A fellow named John Harrison rose to the challenge. He wasn't an astronomer. He wasn't a navigator. He wasn't a cartographer. He was a 23-year-old clockmaker, who produced a timepiece accurate enough and sturdy enough for harsh sea conditions that sailors could compare the local time (based on sunrise) to Greenwich Mean Time and thus get a fix on east-west position. It took Harrison a while to come up with an economically practical prototype, and some more time for testing at sea, but after a mere 59 years, Parliament gave him his money.

Three centuries later, GPS has only just displaced the technology. But the strategy for crowdsourcing invention itself has been rediscovered. It is exactly the founding principle of InnoCentive, a web community of more than 165,000 inquisitive souls in 165 countries competing, like the Netflix prize aspirants, to solve scientific and technological challenges. It's like *The Apprentice*, minus the douchebag host. Or, put another way, what Wikipedia is to cataloguing all human knowledge to date, Inno-Centive is to finding out new stuff.

"Genius can come from anywhere," says Dwayne Spradlin, InnoCentive CEO. "You don't ask someone to solve your problem. You ask everyone to solve your problem."

InnoCentive was born as an Eli Lilly experiment in expanding its organic-chemistry research reach, and later spun off. Lilly remains a heavy user on matters of synthetic chemistry, general chemistry and statistics. Bret Huff, executive director of chemical product research and development there, cites a recent example: the search for an organic solvent more environmentally friendly than the widely used methylene chloride.

"It is generally a very useful solvent for many chemical reactions," Huff says. "However, it is also quite toxic and environmentally unacceptable. Through InnoCentive, we asked the question about substitutes for methylene chloride that would provide all of the benefits of the solvent without the liabilities."

So it floated the project to InnoCentive's network of 200,000 "solvers." "We received about 40 potential solutions from blinded solvers. All of the potential solutions were evaluated by members of our scientific staff. Some of the solutions were quite novel, but not so practical, other solutions were well-known to us, and a few solutions were interesting enough to pursue further."

One of them—2-methyl tetrahydrofuran—has become a high-percentage substitute for methylene chloride.

For most of industrial history, the idea of opening up to the world at large was anathema, the very opposite of the nigh-unto-paranoic trade secrecy that informed every single corporate endeavor. Ask a Procter & Gamble employee about the weather in Cincinnati and you'll probably be told, "Sorry, that's proprietary information." But the cost of R&D as a percentage of revenues has become so high, that even the most compulsively controlling organizations are beginning to exploit digital-age opportunity to cast a wide net for scientific and technological solutions. P&G has been an early and enthusiastic user of InnoCentive's "solvers network." For example, it turned to InnoCentive when its in-house team was stymied on a dishwashing-liquid improvement, one that would turn the dishwater blue once sufficient soap had been added. This was taken up by a homespun Italian chemist, who created a new dye that did the trick and won $30,000 for her trouble. Thus has the notoriously secretive company begun gradually to open up, like a sea clam or China. (Although not *entirely* open up. Given an opportunity to boast about the InnoCentive partnership, P&G dispatched an executive to politely decline.) Any way you shake it, these are remarkable steps. Whether it's P&G, Eli Lilly, DuPont or Boeing corporate culture is changing.

"We've been a monolithic command-and-control culture," Spradlin says of corporate R&D—the long history of which he describes as: " 'I'm going to hire all the best PhDs from Stanford. Period.' "

But if cost has been the catalyst for fanning out beyond a small cadre of usual suspects, the bonus is that solutions often don't even come from the usual disciplines. One InnoCentive client, for instance, was the Oil Spill Recovery Institute, which for two decades has been trying to clean up Alaska's Prince William Sound from the environmental disaster of the Exxon Valdez. The water is so cold, and the 80,000 gallons of crude oil so viscous, that all the king's pumps and all the king's pipes couldn't put the ecosystem together again. It was a problem that defied the best mechanical engineers and the best petroleum engineers and the best hydraulic engineers and the best chemists tasked to solve it.

"It turns out that the solution came from a construction engineer from the Midwest," Spradlin says. "What he recognized was that keeping oil liquid and fluid was very much like keeping cement liquid while pouring a foundation. His solution was to use off-the-shelf vibrating equipment and vibrate the oil so it could go through the pumping facility."

In the winter of 2009, testing is scheduled to commence.

Open Souse Innovation

All of the above are classically digital-age examples of Listenomics. But there is no rule saying all exercises in listening must take the same form. Even the sandwich 'n' blather focus group — if not mistaken for actual data, as we shall soon discuss — can be a useful tool. And there is also no rule that Listenomics must be exercised at every stage of the product life cycle. For proof of this, we must head down under.

If you've seen the Foster's Lager commercials, you might have some, shall we say, stereotypical ideas about Australia. You might think that Aussies are constantly in hand to hand combat with sharks and croco-diles, or boomeranging, or pulled over to the side of dusty dirt roads in their Range Rovers, dislodging kangaroos from their grilles. These are all slight exaggerations. In fact, 95% of the population lives on the crescent of southern coast in Sydney, Melbourne, Brisbane, Adelaide or Perth, cities that are by and large croc- and roo-free. Not only are most Aussies urban and unarmed, it turns out that Foster's isn't even really "Australian for beer." It's an export brand with a very low domestic pro-file, apart from sneering disdain. Basically, Lion Nathan Ltd.'s Toohey's is Australian for beer — at least in New South Wales. In Melbourne, Carlton & United's Victoria Bitter is Victorian for beer. In Adelaide,

Lion Nathan's West End is South Australian for beer. Lion Nathan and Carlton & United's breweries between them utterly dominate the market, using their distribution muscle to control nearly every beer tap in every pub in the country. They have so locked up the market that you'd have to be nuts, or drunk yourself, to enter the fray.

Aha! Welcome to "The only beer company Built by the People for the People." Welcome to Brewtopia — Australian for weird.

If you can find it. The global headquarters of Brewtopia is unit #3 in a small commercial park in the Sydney suburb of Gladesville. It consists of about 1200 square feet of office space above about 1200 square feet of poured-concrete garage. The micro-brewery and the warehouse are situated a bit farther out of town, but 46 – 48 Buffalo Road is the nerve center, teeming, during my visit, with nearly nine employees. Downstairs, amid stacked crates of naked green longnecks, three workers affix custom labels to bottles at a less than breakneck pace. Upstairs, in a cheerful but nondescript office painted in corporate yellow and blue, the mood is also laid back. There are opium dens exhibiting a greater sense of urgency. Here, in his windowless back office decorated with autographed rugby jerseys, CEO Liam Mulhall recalls the heady days of 2002 when he blundered into the beer business.

It is not a pretty tale.

"Myself and a couple of guys," he says, "decided for various reasons to bring a beer to market. The various reasons were we were horribly drunk on a golf course." And their business plan, such as it was, wasn't so much an entrepreneurial idea as a dirty joke expanded into a piece of conceptual art. They contrived to develop a beer called Blowfly, nominally named after the indigenous insect, in order that one day men all over Australia could walk up to the bartenders and bark "Give me a Blowie," which is colloquially pornographic in the obvious way. The thought of it, fueled by an overdose of Toohey's, just cracked them up. That's not surprising. What's incredible is they followed through.

Their idea presented only five obstacles 1) Australia's impregnable beer duopoly, 2) none of them had any idea how to brew beer, 3) or knew anything about the beverage business, 4) or had any money to speak of, and 5) on the face of it, trying to build a brand around a hypothetical double-entendre is stupid. But Liam Mulhall and his mates were not your typical drunken golfers. They were drunken golfers with stick-to-it-iveness, and they were inspired. As their scheme envisioned no particular style of beer — lager, pilsner, stout, ale — nor any other brand characteristic save vaguely disreputable name, it occurred to them that they could convert their vast fecklessness and ignorance into a benefit.

Mulhall, a salesman for Red Hat Linux, had made a living selling applications for open-source software. Why not market Blowfly as an open-source beer? But rather than beginning with an established business and incorporating Listenomics into the previously hidebound top-down corporate culture, these guys would create a brewtopia, building the company from the very outset on the accumulated wisdom of the masses. Over a period of 13 weeks, they opened up their product to public discussion and awarded actual shares of stock to anybody who weighed in on lager versus pilsner, green-bottle versus brown bottle and so on. Granted, the exercise was more gimmick than business model ("I'd sit there and go, 'Oh, no. What am I going to ask this week?'" Mulhall recalls.) but it was a gimmick that gave the power to the people. "The community kept voting. At the end of 3 months we had 15,000 members." Yes, unaware of the adolescent stunt into which they had been lured, 15,000 brewtopians coalesced around the world in service of Blowfly.

The final recipe:

> ***Beer Style:*** European Lager
> ***Color:*** Straw
> ***Hops:*** Saaz & Super Alpha Hops (Imported from Germany)
> ***Malt:*** Australian Pale Malt (Geelong)
> ***Nose:*** Earthy fruity aromas
> ***Palate:*** Low bitterness, smooth on the mid-palate
> ***Alcohol Vol:*** 4.5%

The final name: Blowfly. In a harbinger of things to come, this was not put to the brewtopians.

Listen, Is That Opportunity Knocking?

Though the partners were beverage-business novices, they did have solid instincts for self-promotion, managing to convert a comedy of errors into a kind of reality show that captured a lot of publicity and, briefly, the imagination of Sydney beer drinkers. Alas, as a bona fide competitor, the Blowie wasn't long for the world. In short order, the negligible distribution and cost of subcontracting the actual brewing conspired to choke profitability. The noble experiment — in golfing terms — was falling apart on the back nine. But then the phone rang. Someone named Reg, who happened to be nicknamed "The Blowfly," was having a birthday. Could the boys slap some custom labels on a case of Blowies as a special treat?

No worries, mate, they said. Reg got a big thrill. Not only that, he told his mates at his workplace about it, and Reg happened to work at an advertising agency. Next thing, one of the agency's clients wanted some beer with custom labels. Ten thousand cases of it. And very soon, Brewtopia was doing 5,000 – 9,000 cases per month not of Blowfly, but of Larry's Beer, Bullseye Rocks Brew, Michele & Lewis's wedding day beer and any other label that paying customers might wish to personalize — including names far dirtier than Blowfly. Brewtopia sells a case of Blowie here and there, but it is now almost exclusively a custom bottler of beer, wine and water, for individuals, corporate promotions (Citibank, Amex, Yahoo!, Paramount Pictures, the Foo Fighters) and restaurant private labels.

Also, management now decides everything. Brewtopians at large have lost their voice.

"We used the great unwashed to a point in time," Mulhall shrugs. "I'd like to implement some of the things people suggest, but they're not always in the interest of the shareholders. There is 'open source,' and there is 'making money.' If you left it up to the people, you'd be giving beer away for free."

Point taken, and it may be that the principles of Listenomics do not necessarily apply once a fixed, fairly immutable product is brought to market. On the other hand, you can argue that a custom-beer business, in which every order is based specifically and entirely on the desires of the individual customer, is the ultimate expression of power in the hands of the consumer. But, as I was saying a few pages back, there is more than one way to exploit consumer involvement, and there is surely more than one way to involve them. And the internet needn't always be the means. Ladies and gentlemen, this is about the Ostomy Roadshow 2003.

The Customer is Always Right. Colon.

Okay, the Ice Capades it wasn't. But a splendid example of Listenomics it was, sponsored by a Danish medical-supplies company called Coloplast. (Yes, Lego is also Danish, which not only creates a nice symmetry for this chapter, but also makes you wonder if Scandinavians are better listeners than the rest of us. Although, you know, Volvo was about a decade of American nagging late with cup holders, and Torvald, of *A Doll's House* fame, well, he never heard a word poor Nora said.) The Ostomy Roadshow was something like a mobile focus group — a bus traveling around the U.K. to connect with those, to paraphrase the old Cook and Moore routine, deficient in the colon area to the tune of one. The idea was to get a better understanding of their problems and their needs.

It is, of course, not easy to be an ostomy patient. It's awkward, sometimes embarrassing and not infrequently exacerbated by unpleasant complications. One of them is hernia, a side-effect of surgery involving a rupture of the abdominal wall, through which intestines may protrude. Because surgery is seldom an option for these patients, for decades the standard treatment for such hernias — indeed, the only treatment — was a truss, a set of belts and spring-loaded pads that compress the abdominal bulge. Needless to say, the uncomfortable contraption adds insult to injury. What Coloplast learned during the Ostomy Road Show, according to Copenhagen marketing consultant Soren Merit, is that patients dreamed of another solution.

Probably, hernias don't figure much in your dreams. Probably, they aren't even on your radar. Nor were they for Coloplast, which was doing fine in the colostomy bag business and deficient in the hernia-awareness area to the tune of a lot. The visits with patients were therefore a revelation. As Merit observes, to have an ostomy hernia is to think about it "Twenty-four hours a day for maybe 20 years. For them it's a high-interest area."

The result, two years later, was a new Coloplast product: Corsinel, sort of industrial strength control-top underwear that replaces the truss altogether. "They have a systematic way of always having a dialogue," he says. "That's one thing. The other thing is they actually listen. They involved patients in the actual process of development."

Now, you might ask, "It was a mobile consumer-research unit. Just a focus group on wheels. What's so Next Big Thing about that?" Not a bad question, either. But you have unwittingly gotten to the heart of everything wrong with the bad old days and everything promising about the Brave New World.

For starters, Corsinel didn't just ask a few questions, hand everybody 50 bucks and send them on their way. This was a concerted effort to understand the day-to-day challenges, frustrations and desires of the consumer. Upon learning hernia is an ongoing nuisance, they could have easily said "Thanks for your time. We'll handle it from here." But they didn't. They kept the same consumers involved in every stage of the product development. More significantly still, they actually understood what a focus group is for. Which is more than you can say for almost everyone else in the world who uses them.

No matter how carefully you select the six or eight people around that conference table, and no matter how skilled the moderator, what you harvest at a focus group is not data. Let me repeat that: focus group results are not data. They are gab. There is not one single result you obtain there that is projectable against a population larger than the one in the room itself (and maybe not even them, because group dynamics dictate that people convened in such circumstances don't necessarily say — or even know — what they think about the subject at hand).

The technique is popular and the industry is large not because it is in any way reliable. On the contrary, the "results" are statistically as likely to reflect the opposite of reality as they are to reflect truth. Eight people do not constitute a sample. They barely constitute a cocktail party. But focus groups flourish for two reasons:

They're cheap — pennies on the dollar versus actual research conducted in privacy with a statistically significant sample of respondents and a low margin of error.

They make you feel *sooooo* good. You can sit there on the client side of the two-way mirror and delude yourself into thinking you're getting in touch with the consumer, or the electorate or whomever. This is an especially splendid feeling when the conversation seems to validate the decisions you've probably already made. "You see, Frank? These people do want tort reform/yogurt shampoo!"

The pity of this foolishness is that the true benefit of the focus group is squandered. What they are good for, what they are actually invaluable for, is mining insights. In other words, for revealing something you hadn't thought of. That's why you should, for instance, test TV commercials with focus groups: not to generate some sort of bogus test score that supposedly tells you if the thing communicates the copy points, but in case someone in the room says: "I don't think you should be doing homophobia jokes in a Super Bowl candy bar ad because I don't want to explain it to my kids and I think it's mean and it actually kind of grosses me out." Yes, sometimes in the corporate bunker the obvious can escape you. In the some back room at the shopping mall, with the help of some random dullards dragged from in front of Kay Jewelers, the obvious can smack you right upside the head.

Think of Sgt. Joe Friday. Episode after *Dragnet* episode, he and Gannon would interview citizens and ask them questions about the crime, only to come up dry. Then, invariably, in response to no question at all, the citizen would chime in with an utterly unsolicited parting comment.

"Oh, Sergeant, this probably isn't important, but I did see a man running from the crime scene."

"Was he saying anything, ma'am?"

"Well, nothing that made sense, Sergeant. I was quite a distance away, but he seemed to be complaining about his truss."

Chapter 9

OFF, OFF, OFF MADISON

THIS CHAPTER, LIKE THE LAST, is about opening things up. It's about rethinking notions of proprietary control and expertise. It's about casting far and wide for solutions to matters that have historically been the purview only of in-house specialists and a cadre of trusted suppliers bound by exclusive contracts and deathly fear of losing a client.

In short, it's about tapping into the infinite resources of The Crowd. Please note, however: not as an end in itself.

With so many institutions of mass media and marketing crumbling before our eyes, it's easy to assume that they can all be re-imagined online, using approximately the same set of digital tools. Alas, neither ready technology nor eager participants means guaranteed of success. Shortly I'll introduce you to a company called XLNTads.com with a bold scheme for crowdsourcing the production of TV commercials to a vast pool of talented and not-so-talented amateurs equipped with tools available, for chump change, at Best Buy. To the investors in the enterprise, it seems like such a no-brainer.

Well, why not? Give or take a few details, it's been done before.

For instance — courtesy of author Jeff Howe, coiner of the term "crowdsourcing" and author of a wonderful book of the same name — consider a company called iStockphoto that has precisely used technology to "democratize" a hitherto closed universe of visual production. Actually, for this company, "scheme" isn't really the best word, because it suggests premeditation. Rather, iStockphoto began as an online photo-sharing community in which visitors who uploaded their stills (or video or animation) were free to download an equal number at no cost. Only when the site's popularity began to make it expensive to host did it begin charging a nominal sum for images. Then a higher nominal charge. Today, from an archive of 12 million items, the price ranges from a $1 for a basic still photo $86 for an HD video file — which is still a fraction of the licensing fees commanded by traditional stock houses. That includes Getty Images, the industry gorilla, which responded to the threat posed by iStockphoto by acquiring it for $50 million. The rationale: if the business paradigm shifts it's best to have your palm on the shifter.

One of the lessons of iStockphoto is that while the digital production tools have the power to decimate mature industries, they also have the power to create new ones. This was precisely the thought of Ross Kimbarovsky, a restless intellectual-property litigator, and Michael Samson, a restless ex-Hollywood executive, who in 2006 had a notion to outsource digital video production to south Asia, where video editors — like sewing machine operators and call-center employees — cost 2/3 less than their American counterparts. In scouting the marketplace, Samson recalls, they came upon an online forum for graphic-design students in Malaysia.

"They were competing against each other — not for money. It was a club, basically. And the work was phenomenal. We couldn't believe it. I said, 'This could be on the side of a bus in New York City.'"

And high-priced New York graphic designers could be thrown underneath the same bus. Having followed the fortunes of iStockphoto, Samson and Kimbarovsky began thinking about online means of opening-up the design marketplace to "a huge pool of providers," and pitting them against one another project by project. The Malaysian students competed out of pride, but what if winning submissions were rewarded with cash money?

Thus crowdSPRING.com was born.

I Did Say This is About Exploiting the Digital Revolution, Right?

I got my book cover there for $500. A Walla Walla, Washington dentist got a logo there for $900. The Amazon Foundation got one for $333. And at about the time those projects were finished, as my *Advertising Age* colleague Rupal Parekh reported, Pepsi updated its logo. Pricetag: $1 million.

The slight disparity between The Old Model and crowdSPRING's new one — in this case, $999,667 — is at the root of what Mike Samson calls "pushback from the traditional design community." It is obviously no fun for those folks to see students, hobbyists, dilettantes, hungry freelancers and undocumented aliens not only working for peanuts, but working on spec — a practice deemed at best self-defeating and at worst economic slavery. When *Forbes* did a fairly glowing article about crowdSPRING, the online comments were scathing, along the lines of early-19th century craftsmen displaced by steam-powered factories.

> *That's the great thing about the internet. Knowledge,Experience, research and expertise don't really matter anymore. All you need is an unlimited pool of people with mediocre ability who are hungry enough to dash after the last scrap of food, dangled by a greedy ignoramus like crowdSPRING.*

Another commenter sought to explain why a superficially attractive entry from an untrained amateur cannot measure up against the work of a trained professional:

> *It's called research and strategy, you buffoon. crowdSPRING cons high school kids who can 'draw pictures' into doing free sweat-shop work. I can use Excel; does that make me an accountant? A company's logo should be a reflection of their brand. It's not just a haphazard doodle … Otherwise you've reduced design to nothing more than window dressing.*

It is true that high-end design shops conduct research and testing to maximize the tonality and communicative strength of logos and trade dress of all kinds. But none of that stuff is magic and a whole lot of it is bullshit — in some cases $999,667 worth. Ultimately, a design is either good or it isn't, irrespective of its provenance. The Wright Brothers were bicycle mechanics. Benjamin Franklin was a printer. Alexander Borodin was a chemist. Think he was a charlatan? Tell that to the Polovetsian Dances and the Symphony No. 1 in E flat major.

The more compelling issue is whether this method of democratizing an industry is inherently unethical. In an average crowdSPRING project, one winner is compensated, shall we say, unextravagantly and 67 work for free. Even if you accept that to be an inevitability of the digital revolution, doesn't it have a sort of *Oliver Twist* vibe? Am I not an exploiter of the designer of my book cover and, especially, of the 100+ who submitted and failed?

Unsurprisingly, Samson doesn't see it that way.

"Our position is, we truly believe we're expanding the market," he says. "These [buyers] are people who did not have access to graphic design before. They could draw their own logo. They could hire their friend's niece. They could go down to OfficeMax and pick some crappy clip-art out of a catalogue."

Or they could go to the highly-trained professionals at an established graphic-design house and plunk down several grand for a retainer, while simultaneously submitting their logotype fate to the vision of one soul who may, for all his or her training, be uninspired by the project or maybe kind of suck. Granted, plenty of the submissions for this book's cover were not too impressive and quite a few were awful. But I still had a tough time choosing from among the two or three dozen that were — in my professional opinion as an advertising critic — pretty damn good. You can check three of them out at TheChaosScenario.net.

As for economic dislocation, on that point Samson cannot disagree. He can, however, shrug it off. Technology and labor costs always have imposed disruption, and always will.

Never mind the Industrial Revolution. "If you were a mill worker in New England in the 19th century when the textile industry was moving to North Carolina, you were threatened. If you were a mill worker in North Carolina in the 1980s and the industry was moving to Vietnam, you were threatened. We realize we represent something scary to a lot of people."

Because revolutions do.

Which brings us at last to XLNTads.com.

Hot Beverages, Cold Market

I first visited that company in the spring of 2007, another journey in my ongoing Travelogue of Metaphors Almost Too Good to be True. For where was this startup situated but Conshohocken, Pennsylvania, a poster child for life after economic dislocation? Once upon a time, Conshohocken itself had been a thriving mill town. Then, one by one, literally and figuratively, its industries went south. For most of the past four decades, it's been a moldering ruin on the western outskirts of Philadelphia, notable for its abandoned factories, overgrown rail sidings and rusty water towers. These days still it is mainly a former mill town, dotted with those oxidized artifacts of the long-departed manufacturing economy, but also suddenly thick with office condos, cookie-cutter business hotels and low-slung light-industrial parks. It isn't the smokestack economy it once was, but it has gradually adapted. Conshohocken's biggest asset is proximity to the crossroads of I-476 and the Schuylkill Expressway, a mile from which — in a nondescript small-business complex — I found the offices of XLNTads.com.

The walls were beige, the carpets were industrial. The contemporary artwork was not from Sotheby's but from the local Marshall's, and the loading dock was the floor next to the front door. A few weeks after the move-in, cartons of copy paper were stacked pretty much exactly where you'd expect to see a potted fichus.

On the other hand, observed acting CEO Neil Perry, "Not a lot of start ups have a coffee maker *and* a hot-water dispenser, so ..."

Kidding. The man was kidding. The guy wasn't really boasting about instant hot water. He was being self-deprecating, because he's a funny guy. Which is good. He was going to need a sense of humor trying to get XLNTads.com off the ground. As we shall soon see, this enterprise was by no means a sure thing. On the other hand, Perry and his principal partner, entrepreneur Rick Parkhill, were on to something with a lot of buzz to it — and that something was consumer-generated advertising.

Well, why not try to make a buck on CGA? It was certainly the trend *du jour*, having been so widely talked about at nothing less than the 2007 Super Bowl. There, in the midst of the highest-profile advertising event of the year, were three commercials either inspired by, written by or wholly produced by consumers. The best of the lot, as it happened, was the least consumer-generated — a simple, bittersweet, Basin Street elegy for the departed football season. As Louis Armstrong's horn wailed plaintively in the background, we saw all manner of fans representing a cross-section of America, mourning a season that ended too soon. We also saw the embodiment of a modest human truth, equal parts embarrassing and poignant: We care far too much about a stupid game. The spot was shot by Joe Pytka, the most accomplished director in the history of television advertising, but it was inspired by an ordinary fan who submitted the idea as part of an NFL promotion.

Another spot was also professionally produced, by Chevrolet's ad agency, Campbell-Ewald, for the car-truck mutant called the Chevy HHR. This one was written and storyboarded by students at the University of Wisconsin-Milwaukee. Hard to know where things went wrong — at the college or at the ad agency — but the scene of guys spontaneously throwing off their shirts and getting all erotic on a passing HHR was completely incomprehensible. About $2 million worth of incomprehensible.

The best of the three could and should have been an entirely consumer-generated spot, for Doritos. Alas, speaking of incomprehensible, online voters in a Frito-Lay contest failed to vote for the best of five finalists, or even the second best. Or the third. The satisfactory-but-unillustrious winner dramatized a goofy encounter between star-crossed lovers, in which every aspect of the episode is described in onscreen type that also happens to describe Doritos attributes: "bold," "cheesy," etc. It wasn't horrible, but it paled miserably next to the unvictorious finalist titled "Mousetrap," which showed a guy leaving a Nacho Cheese Dorito next to a baseboard hole to bait an annoying household pest. He sits backs to enjoy the rest of the bag as he waits. A moment later, a giant, man-size mouse comes crashing through the wall to beat the crap out of him and take his Doritos. Hilarious. Should've won hands down.

But, let's face it: the point of none of these ads was to find the next Apple "1984." The point was to run a successful consumer promotion. A gimmick, in other words. The bonus is that — except for the Chevy HHR thing, which totally, you know, reeked — the stuff wasn't too bad. Meanwhile, the "consumer-generated" concept generated lots of consumer interest, and even more publicity. This perhaps emboldened other advertisers, in varying degrees, to jump on the CGA bandwagon. Despite the HHR fiasco, such advertisers as MasterCard, JetBlue, Converse, Heinz and General Electric all seem themselves to have absorbed the meaning of YouTube, embraced the million-monkeys way of looking at things and recalculated the most efficient way of producing a useful concept — namely: not necessarily by bankrolling pricey "creatives" in $100/square foot Madison Avenue offices. That calculus, of course, was the *raison d'être* of XLNTads.com.

"What's the harm?" asked Perry, a veteran of McDonald's brand management who left consulting to take this gig. "And what if going through the exercise you come up with a great idea?"

The answer to that, obviously, is there is hardly any exercise not worth going through if the reward is a great idea. But Perry wasn't necessarily posing the right question. Presently, I'll propose a better one. In fact, I'll suggest several. For the moment, though, it's easy to see why CGA then and now continues to hold an attraction. For starters, as the Super Bowl amply demonstrated, it's genuinely a *zeitgeist* enterprise. Bloggers, vloggers, and short-form video artists are changing the world of content by stuffing the digital pipeline with work of their own creation. Remember what media guru Rishad Tobaccowala said back in Chapter 3: "If you aren't posting, you don't exist. People say, 'I post, therefore I am.'"

Once again, Rishad is not wrong. Increasingly, e-citizens are taking YouTube at its word. "Broadcast yourself" is one slogan the world has taken to heart and, in YouTube's short-form universe, the 30-second spot is an inviting genre. They've enjoyed/endured commercials their entire lives. They (think they) know how ads are constructed. Thirty or sixty seconds are bite-sized chunks, made to order for the short-form universe.

Another benefit is cost. Compared to, say, $350,000 for an ordinary network TV spot, a CGA commercial comes in for pennies. A lot of this has to do with production values (It is impossible not to tell the difference between Madison Avenue slickness, however superficial, and CGA rough edges). But it also has to do with the nature of remuneration, and what exactly consumer ad generators are seeking.

"These people are amazing," Perry gushed. "We kept probing about compensation. They said, 'It's not that. It's watching it get posted and seeing how many hits it gets.' They're hungry creatives, but they're hungry in a different way. They're hungry for recognition."

Which is free. That alone should give the agency world pause. As previously noted, "free" is a difficult price to compete against.

Furthermore, even in the most dedicated ad agency, such enthusiasm is difficult to replicate. The main reason marketers have experimented with CGA so far has been to cultivate exactly such enthusiasm, turning amateur ad people into brand ambassadors, or viral vectors, or birds and bees pollinating the meadows of the marketplace. Choose the metaphor you prefer. No marketing message generated by a company itself offers the credibility of a message from a (financially) disinterested civilian.

I Advertise iPod

Consider one of the pioneers, George Masters.

In 1990, a Welsh band called the Darling Buds released a song called "Tiny Machine." After 700 weekends, Casey Kasem still hadn't played it on the radio, but, no matter, George Masters had it on his iPod Mini. One fine California day in June 2004, Masters was on his way to work, listening to "Tiny Machine" on his tiny machine, when inspiration struck.

"I thought that would be a great jumping off point for creating an advertisement," he recalls.

His assessment would prove to be correct, but there was an obvious problem. Masters didn't work at Apple in Cupertino or TBWA/Chiat/Day in Playa del Rey. He worked at a vocational high school in Orange County, which, suffice to say, didn't have the iPod account. Masters, however, was undeterred. A video artist by training, he sat down at his Mac and started crafting an iPod commercial. It took him five months to complete, but when he was done he posted it online. Kaleidoscopic and pastel-laden, it looks nothing like TBWA's dancing Day-Glo silhouettes. But Masters' piece is a masterpiece all its own, and it quickly went viral. Very viral. Millions of people viewed an iPod ad that nobody paid for, because, in the words of author J.D. Lasica, "Something that's genuine, true and authentic, even if it has a commercial message in it, is going to resonate."

And that native authenticity is out there, like Arctic oil, just waiting to be tapped. Pitiful as this may sound, there are people all across this great nation of ours who give immense amounts of thought to, for instance, consumer electronics. They're not in it for the money, either. They just plain care.

"It was never my intention to profit from it," Masters told me. "I still haven't profited one dime from it. In fact, it's cost me money."

So why would a 37-year-old man invest a half year of free time to advertise somebody else's business? Masters answers the question with a question: Why does anyone devote time to the things he is passionate about?

"There's some guy in his garage who's been working on a hot rod for three years. Andy Warhol painted soup cans, right? Guy loved soup."

Or maybe Warhol's motivation was a bit more ironic, but George

Masters was undoubtedly smitten with his iPod, which made his video less of an ad and more of a love letter. Once again, what is more poignant and genuine than a love letter? (Hint: Nothing generated by Omnicom.)

Monkey See, Monkey Do

Of course, not every consumer ad-generator is George Masters, and surely not every CGA is "Tiny Machine." MasterCard learned that only too well in 2007 after staging a promotion in which consumers were solicited to fill in the blanks leading to the brand buzzword "Priceless." For the company, which has publicly discussed its determination to reduce advertising production costs as a ratio of its overall ad budget, success in this experiment could have been priceless, too. Never mind tapping the enormous value of consumer evangelism, an agency spot costing $350,000 is $350,000 more expensive than one costing nothing. Alas, no such luck. In a June 2007 speech to the Internet Advertising Bureau, Cheryl Guerin, MasterCard International's VP-promotions and interactive, reported "we were hard-pressed to find a lot of good ads."

And I'm, like, duhhhh. Even in 2009, most consumer-generated ads aren't the stuff of undiscovered geniuses mining their intimate personal experiences for rare masterpieces like "Tiny Machine." On the contrary, mostly CGA has been the stuff of tiny little talents with tiny little budgets pursuing tiny little ideas. Sadly, citizen "creatives" tend simply to be amateurish versions of professional creatives. Rather than coming at advertising solutions from utterly new directions, they palely mimic the techniques, conventions and clichés of Madison Avenue.

This phenomenon became all too apparent in 2006 when Current TV, the digital cable channel devoted to consumer-generated content, persuaded Sony, L'Oreal and Toyota to participate in a CGA contest. They called it "V-CAM," Viewer-Created Advertising Messages. They might just have easily named it V-CRAP. My God, were the entries awful. The majority were two-bit exercises in homemade digital effects. The remainder were vignettes starring the filmmakers, their friends or their kids. As a group, they made *Funniest Home Videos* look like *Wings of Desire*. That the submissions mainly lacked technical sophistication was no surprise. What was striking was the utter vacuum of underlying ideas. For lovers of the status quo, this is encouraging, because even the bottom tier of agency-produced ads worldwide usually have a point to them. It's often a stupid one, but it's there. In the V-CAM contest, it apparently never occurred to a single entrant that the video was supposed to make the viewer more favorably disposed to the product. The only thing that did seem to occur to the entrants, in fact, was "Hey! Look at me!" Like the man says, "Broadcast Yourself."

Not long ago, I got an email from a friend, who was writing to me not as an ad critic but as one of the 2 or 3 trillion people on her contact list. Here's what she had to say:

Hello to everyone I've ever known! Long story short, there is a Heinz Ketchup commercial contest going on and my children are in one of the entries. It is up on youtube today and it is very funny. The filmmakers are young, energetic and very creative. If you have a second, here is the link. Take a peek at it and if you are so inclined you can rate it and make a comment. The top fifteen chosen by Heinz will be voted on by you the public and narrowed down to five. Those five go on national TV and are voted on again, by you, the public. The top commercial gets 57,000 smackers. Anyway, it was a fun project to be a part of and I want to do all I can to support these guys. Thanks for watching!!!!!

So I watched. The scene is a child's birthday party, with three dorky folksters singing the kids half to sleep in the back yard till "Ketchup Man" — a man-size squirt bottle — crashes through the fence and saves the day. He turns the nerdy trio into hard rockers, he exchanges the little girls in the wading pool for post-pubescent girls (to the delight of the pre-pubescent little boy between them), and transforms Dad at the grill into a red-faced devil, too.

Playing Ketch-up

So, yeah, got it: "Heinz ketchup turns a dull party into fun." You can't miss the selling message — although that message isn't, uh, true. You've got to hand it to these energetic film makers that they even bothered to assert a selling premise. The slight problem, from a branding point of view, is the particulars: the fun comes courtesy of Satan. Ketchup Man has a red face and diabolical eyes. He accomplishes his magic by means of bitch-slapping his adversaries with a Spiderman-like stream of condiment. And once the party is over, presumably Dad continues to be Lucifer.

Not to nitpick or anything, but can you see how this commercial, were it to be part of the company's ad campaign, might trigger some pushback? There are a lot of people in the United States who believe literally in the devil, and my guess is they consume a disproportionate amount of ketchup. They don't even let their kids dress up for Halloween, because they think it's satanic. How long do you suppose they'd wait to begin the Heinz boycott? (Never even mind what the clown was doing before Ketchup Man arrived, except that it involved a balloon being inflated from the area of his groin and a little girl running away screaming. Hey, who doesn't love a little pedophile humor in a ketchup commercial?)

Once again, there's plenty of that in the agency world, too. Whole books have been written on the subject. (*And Now a few Words from Me* by Bob Garfield, McGraw-Hill 2003. Christmas is just around the corner. Order now!) But take my word for it — and I know this is damning Madison Avenue with faint praise — professional vacuousness is better than amateur vacuousness any day. And even if the amateur work manages to approximate professional production values, if that's all it offers, what has been gained?

Consider the winning V-CAM spot, called "Transformation" from a 19-year-old Minneapolis animator named Tyson Ibele. Using logos and stills provided by the advertiser, Ibele depicted products morphing — a la Transformer toys — into one another: a boom box into a plasma TV into a notebook computer into a camera into a handheld Mpeg player into the Sony logo. Meanwhile, on-screen type punctuated the images: "Innovation. Compact design. Experience. Vision. Adventure. On the move. Groundbreaking. Revolutionary."

At the time at least, Ibele was animating for a small Twin Cities digital house, so its no wonder the piece was well-constructed and the effect seamless. It's also no wonder that the idea came to him; it's been done quite a bit before, including a wildly overrated Renault spot that won a lot of international awards while selling, I'm pretty sure, zero Renaults. Like that expensive exercise in "creative" self-involvement, "Transformation" transforms for its own sake, not to make any positive statement about the brand.

Which, come to think of it, doesn't suggest much transformation at all. What's the point of having a revolution if the revolutionaries just imitate the old order?

That, in fact, is one of the questions that Neil Perry at XLNTads.com might in 2007 have been thinking about. And here are a couple more: Why antagonize your agency or agencies — or publicly humiliate them, actually — by soliciting advertising ideas from amateurs, including middle schoolers, morons and perverts? (Yes, there is some value to sending a wake-up call to the old guard, but it's worth remembering that some people don't like being rudely awakened. Remember the rage of the graphic-design community at the audacity of crowdSPRING to bypass them?)

That's one thing. Another is the very real possibility of being shot with your own gun. Who can forget what happened in 2006, when Chevrolet used an episode of *The Apprentice* to promote chevyapprentice.com and a chance to construct a Chevy Tahoe commercial out of components provided online? No doubt General Motors would have preferred a bunch of paeans to ruggedness and style, off-road capability and payload — that sort of thing. There certainly was some of that. But all the blogosphere and media attention was focused on stuff showing the gigundous Tahoe lumbering around the fragile environment underneath onscreen copy like this:

Don't give a fuck; it's a lifestyle … 'Cause as the icecaps melt I've got cupholders and live in a fantasy world … so as Inuit cultures drown and die I've got sound dampening technology. No … I've got reality dampening technology … Iraqi blood: $3/gallon. Giddyap.

There's more, but I won't repeat it because it's a little bit more negative. Anyway, you get the idea. To its credit, GM left even the most scathing videos on the site, wagering (correctly) that it would win more trust and goodwill by hosting the debate instead of suppressing it. But let's just agree that the promotion was less than the PR bonanza Chevy was looking for.

So, yes, if you're looking for reasons not put your brand messaging in the hands of the *hoi polloi*, the Tahoe fiasco would be pretty near the top of the list. Not the very top, however. The main reason is that Consumer Generated Advertising is still … whaddyacallit? … advertising. As the second chapter of this book I believe made abundantly clear, advertising is not the future of marketing. No, it will not entirely disappear. Yes, it will play a role. But the 30-second mini-movie is not a growth industry. Digital tools and database marketing will open channels of communication, and reservoirs of information, that will relegate display advertising as we've known it to a subsidiary role. Most likely, it will serve as a series of signposts — logos and simple messages pointing users to websites where the real action is. This could happen within five years, certainly within 25. Unlike graphic design, which will never be obsoleted, the CGA industry — if it ever becomes an industry — is at best a short-term enterprise. XLNTads.com, ugenmedia.com, Filmaka, GeniusRocket, Zooppa and others are hollering, "Jump in! The water's fine!" But if you're a marketer, you'd be forgiven for not diving headlong into a draining pool.

Three Reasons to Ignore Ketchup Man

Well, then. Let's recap: The amateur ads will mainly be cheap knock-offs of professional ones. Your ad agency is going to be infuriated with you. The stuff you solicit may hold your brand up to ridicule or worse.

And, most alarming of all, the 30-second spot is an endangered species. Doesn't seem like much of a case for CGA. As I said, I'm not sure Neil Perry was raising the most relevant questions — not against CGA, but also not in favor of it, either. It just so happens there are several perfectly sound reasons XLNTads.com and other fledgling competitors would want to fight over market share for a short-term proposition. In fact, there are three of them:

For one thing, you could have expressed the same skepticism (and many did) about the blockbuster deal for Blockbuster. When Viacom paid $8.4 billion to Wayne Huizenga for his video-rental empire in 1994, the binary code was already on the screen. The distribution of video via VHS tapes, and later DVDs, was a business whose days were already numbered. Video-on-demand was already being experimented with. It was as if Sumner Redstone had read about the Wright Brothers and rushed out to buy a railroad. And, sure enough, in the new millennium Blockbuster has tanked (although thanks more to Netflix DVD-by-mail than digital video-on-demand.) Nonetheless, in the intervening time, the doomed company generated more than $60 billion in sales. In other words, just because a business isn't forever doesn't mean there's no money in it right now. On this point, believe me, I am an expert. I make a living criticizing 30-second TV commercials.

So, over the next five or 10 years, maybe there is some money to be made. That's one thing. The second thing is that the digital world vastly expands the potential marketplace for 30-second ads. While Neil Perry is focusing on the Leading National Advertisers, an ever-lengthening long tail of businesses suddenly has not only the physical ability but the means to avail itself the most powerful marketing weapon ever deployed: the 30-second spot. If XLNTads were ever to steer its efforts in their direction, the opportunities could be much richer. Remember, the post-advertising age won't be entirely barren of commercials, and the long tail could well be — ironically enough — where they most thrive. But even that's not the reason shrewd marketers would do well to consider exploring CGA. Yes, it's true, the best way to generate a Big Idea is to cast far and wide for one. But considering the early returns — that even the widest nets are being hauled in with no ideas at all — the argument, as it were, also holds no water. What trolling for spots most achieves is creating the conditions for listening. The biggest argument for going to consumers and asking them to think deeply about your brand is the experience it gives a marketer in going to consumers and asking them to think about their brands. It's about a mentality (or, as they like to say now "mind-set," not to be confused with "skill set," which is what they used to call "skills"). It's about establishing a template for learning to cede control, and for harnessing the wisdom of the crowd.

The Miracle of Cupertino

This is useful experience to have. In fact, it is invaluable experience to have. In fact, in the New World Order, it is an obligatory experience to have. Perhaps this is why George Masters never had to hire a lawyer. I don't know what was going on in Apple Computer's corporate mind when it didn't sue Masters for his lovely viral serenade to the iPod Mini. I certainly believe, though, that the episode was a turning point in American industrial history. Not the charming spot itself, but what Apple did in response:

Nothing.

The Miracle of Cupertino, this was. No cease-and-desist letter. No trademark-infringement lawsuit. Not even a press release to distance itself from this MP3 groupie — a non-reaction that Apple refused to discuss with me, but which no doubt had the corporation's trademark lawyers apoplectic. Because everybody knows things aren't done that way. Since time immemorial, there is a universal corporate protocol for handling unsolicited ideas, and it isn't silent acquiescence.

"The letter's opened up and somebody starts reading it and says, 'Ah, we have to send this to legal,'" says Carla Michelotti, senior VP-general counsel of the Leo Burnett Co. "And the legal department returns it with great courtesy."

It was ever thus, because nobody wants to be accused of stealing an unsolicited idea, because the submitter can't guarantee ownership of the idea himself, because legal already has its hands full doing due diligence on the product claims, artists rights and other legal ramifications of the agency's own output, and because brand names and trademarks are priceless assets which must be protected from abuse by outsiders, no matter how well-intentioned.

Have a grrrrreat! idea for Tony the Tiger mauling a flamboyant Austrian magician? History declares don't bother.

"I can't move into your backyard and just decide what to do with your landscaping," Michelotti says. "It's trespassing. It's taking somebody else's property."

Quaint as that view may sound in the internet age, when desktop publishers deem "fair use" to mean every single paragraph, song or image they can cut and paste, Michelotti is still accurately summarizing the law. When Viacom filed a $1 billion lawsuit against YouTube for trespassing on intellectual property, it may have been testing the limits of the New Millennium Copyright Act, but by all previous statutes and case law on the issue, the suit was open and shut. Millions of computer owners out there are running digital chopshops, and YouTube is the showroom for the stolen property.

That's why Apple's action — in the form of conspicuous *in*action — was such a watershed. It demonstrated a new calculus for the digital age. By demonstrably failing to be vigilant about its trademarks, it signaled that the enormous value of a scrupulously protected brand may be exceeded by the value of an open-source one. Yes, subsequent attempts to litigate against infringers will be more difficult, but the worldwide Apple audience — already more cult than marketplace — suddenly has been empowered with the privileges of membership. If brand loyalty is important to you, there is no greater asset. On balance sheets, it is called "goodwill." With the Masters episode, Apple, for itself and others, unlocked what may ultimately amount to trillions of dollars worth of goodwill.

"The centralized model is essentially inside out," says James Cherkoff, a London marketing consultant who penned an online manifesto on open-source marketing. "You create all the messages and you send them out. The new model is outside in: What you want to do is receive all the information you can from the outside and incorporate them in the processes of the company. They have to actually open up their own systems and the way they interface with the world."

Semi-Pro to Go

Let's then assume, at least for the moment, that XLNTads.com was well positioned to hold the hands of adventurous marketers as they took their first uncertain steps into the world of Listenomics. I first spoke to acting CEO Neil Perry three months before the official launch of the site for paying customers. Sixty days later — five weeks from launch — he had yet to sign deals with many adventurers.

"You know," he reported, "we still have a very good idea and we still have very few clients. The challenge is, because our site isn't really a site yet, everybody is waiting to see who else is going to be there before jumping in. We're not getting objections to the price ($25,000 per month for a three-month minimum); nobody's objecting to the format. But they're big companies and it's a chunk of money, and any time a corporation is spending a chunk of money, they can't make an expenditure for a totally unknown commodity without eyebrows being raised in the back of the house."

Ah, life in Start-Up World. At five weeks and counting, the one deal signed and sealed is with X-Box Live — although not for CGA. This would be for consumer-generated web content. "We're feeling very comfortable on our sales calls, it's just a question of lockin' and loading. I think the next one we'll get is General Motors."

Although — and I'm just guessing here — probably not for the Chevy Tahoe. Or any other high-profile model, for that matter.

"Our sweet spot seems to be more of the small to mid-tier brands," Perry said. "We began by going after Coke and Diet Pepsi. It's starting to feel like Fanta Orange and Dasani water."

It was also starting to feel like a somewhat different business than was first envisioned. While XLNTads.com would still offer advertisers a distinct venue for CGA competitions with no risk of awkward adjacencies — like, say, pornography or violence — the C in CGA may not be exactly Joe Schmo of Anytown, U.S.A.. It's more likely to be, for example, Kevin Nalty of Doylestown, Pa. Nalty, a YouTube veteran with an online following for his funny short videos, is a new breed of semi-pro. He has enjoyed more than a million views, for instance, of his seminal viral "Farting in Public."

Yes, *the* "Farting in Public." But he has done almost 400 others, many of them quite hilarious, actually, and he is on XLNTads.com's Creator Ad Board, a group of other online video auteurs who consider themselves the AAA of advertising's minor leagues. They're really more like single A, but they clearly are vastly more skilled than the rank amateurs doing most of the submitted-to CGA contests around the world.

"The barriers of entry are so low, there's just tons of garbage," says Nalty, who is known online as Kevin Nalts. He uses the semi-pseudonym presumably to safeguard his employer, Merck, where he is a marketing executive. "I market by day and do videos by night," he says, giving him a lot more bona fides than the average "armchair ad watchers who don't know how to deliver a marketing message." He and his fellow travelers have prevailed upon Neil Perry to alter his strategy: namely, to cultivate the semi-pros.

"You don't need a huge cadre of those to have success," Nalty says. "If you can find a few dozen that know marketing and know the rules of this game, then you get something of value. There are people out there lurking in the shadows with great imaginations."

Maybe so. Nalty has already done ad-like videos for paying customers. In one, for Mentos, he is caught trying to sneak a roll of mints into a movie theater. The thing is, the Mentos roll is about five-feet long and it protrudes pretty glaringly from his trench coat. The early version of the ad was much too long, too, and didn't so much end as just stop, but it's a pretty clever idea — and much crisper in the shortened version he completed after first appearing on my blog. Nalty did another for a website called GPSManiac.com, in which we hear the interior monologue of the female-voiced GPS in his car. As you might guess, she doesn't think much of him. The video is — shall I say — under-funny, but his client thought enough of it to post it on its site, headlined, zeitgeistily: "As Seen on YouTube."

His payment? Somewhere between $1000 and $5000.

Neil Perry's new strategy for XLNTads.com was to recruit 500 Kevin Naltys, not only to secure a ready supply of contest entrants to prime the pump for promotions eventually to be hosted on he site, but also to create a clearinghouse of semi-pro talent for brands that so recoil at high-street prices that they're content to shop at the dollar store. If two heads are better than one, after all, 500 heads should be fantastic.

Trolling for Ideas

This is not quite the paradigm shift it would seem, as tapping the Ingenuity of the Crowd is hardly a new advertising idea. Back in 1999, a creative director at DDB, Chicago, saw a spec reel from a young filmmaker doing a piece of unique sketch comedy about male bonding. The short film by Charles Stone III and his friends was called "Whassup." DDB slapped some Budweiser into the scenario, and it became a Cannes Grand Prix winner, not to mention one of the best beer campaigns ever. The truth is, advertising agencies are all the time borrowing from the culture. That's how Bob & Ray became the Piels Brothers in the 60s, how Max Headroom became a Diet Coke character in the 80s, how the obscure urban-acrobatics sport called parkour keeps popping up now. (Not to mention the 30,000 or so *Close Encounters* homages/ripoffs that have shown up in commercials for 25 years.) What YouTube does, though, is vastly simplify the discovery process. Creative teams can troll for ideas simply by typing in keywords — or by just keeping an eye on the videos that percolate up organically to the most-viewed lists. With or without intermediaries like XLNTads.com, YouTube is a virtual open-mic night for every blocked creative team with a broadband connection.

Consider, for instance, a spot for McDonald's that ran in the New York metropolitan area in the summer of 2007. It's two guys on a street corner, one making beat-box sounds through pursed lips, the other rapping about breaded, deep-fried chicken parts:

> *I'm into nuggets, y'all. I'm into nuggets, y'all.*
> *I'm into nuggets, y'all. I'm into nuggets, y'all.*
> *McNuggets, McNuggets what,*
> *McNuggets, McNuggets what, McNuggets, McNuggets what*
> *Ketchup,and mayo. Ketchup and mayo.*
> *Dip it in barbecue sauce. Dip it in barbecue sauce.*

The roots of this moving lyrical tribute go back to Chicago, a year earlier, when two students in the training program of the Second City improv company were preparing for some sketch comedy onstage. One was Thomas Middleditch, the other Fernando Sosa. "We were backstage, 10 minutes before we go on, and doing hip-hop characters back and forth," Sosa says, "just to goof around." That's when, in recognition of rap's famous ability to document the nitty-gritty of urban life, Middleditch started to rap about Chicken McNuggets. This made them laugh, and when they went onstage, they ditched one of the bits they'd been planning on and launched into McRap. Good move.

"It was awesome," Sosa says. "The audience went crazy."

Their friend Matt Malinski then agreed to make a video of the performance, which he duly posted on YouTube with, he insists, "no expectations, really." Well, maybe not expectations, but certainly high hopes. Recall how *Saturday Night Live* bit players Andy Samberg and Chris Parnell were catapulted into prominence when their "Lazy Sunday" rap about the white-nerd life became a YouTube sensation. As it turned out, the 40-second McNugget rap wasn't quite "Lazy Sunday," downloadwise, but over the next two years it did generate almost 2 million page views. And one of the viewers was Chris Edwards.

"I was dying laughing," says the Arnold, Boston, creative director responsible for McDonald's retail promotions in New York City, "and I was, like, I gotta do something. This is just too good to leave there. So I downloaded it into iMovie and worked on it over the weekend."

Sure enough, he figured out a way to cut out some of the stuff, replace it with title cards, and slap a logo on the end in 30 seconds. That is: the length of a commercial. Then it was just a question of tracking the boys down to inquire as to whether they'd like to be all over New York TV rapping about a fast-food item.

"We were worried they'd be like, 'No, we don't want to look like sell-outs'" Edwards says. "But they were excited, they were very excited to be put on television. They're psyched to get the exposure." That is perhaps an understatement. Sosa himself is a bit less circumspect: "This is fucking retarded. The attention this is getting is ridiculous. It's insane."

It's hard to say what makes the video so compelling. Maybe it's because nobody would rap about McNuggets. Or maybe it's because McNuggets should have been rapped about long ago, because they represent exactly what rap is supposed to explore: the textures of the inner city. Truth be known, the guys themselves can't quite put a finger on it. Because, really, how can you ever? Once, in preparing for Chicago's SketchFest, they spent hour after hour honing a bit about a kid and his friend walking into the house and finding the kid's dad watching porn on TV. They cracked themselves up every time they ran through it, yet nobody in the actual audience was even slightly amused.

"They just would not laugh," Sosa recalls.

Well, this time the audience did laugh. The video was already a mini-cult phenomenon when Edwards ran across it. And it just so happened that the subject matter was a perfect fit for an Extra Value Meal summer promotion in the New York area, where Arnold has regional responsibilities.

"It was kind of a fluke," he says, "but a lucky one at that."

So what will be the upshot of all this serendipity for the hamburger company seeking to exploit it? Edwards thinks it will be a spontaneous, unsolicited flood, on YouTube and elsewhere, of CGA — most of it probably from rappers "thinking "maybe we can get it on TV.'" Which is fine. The key fact at work here is that, indeed, phenomena by their very nature cannot be predicted. For every Nike "Ronaldinho" and Dove "Evolution" that became viral sensations, thousands of other supposed "viral" postings by ad agencies never infected anyone's imagination outside the agency conference room. For sure, the vast YouTube-o-sphere exponentially increases the slim chance for virulence. But the important word in that sentence is "slim."

Getting Where?

Back in Conshohocken, in August 2007, the remnants of a tropical storm had rain coming down in sheets, leaving a reverse archipelago of dreary puddles in the parking lot. This was the view outside of Neil Perry's office window, but no worries. He was growing accustomed to "dreary." It had been nearly five months that he and his sales team had been out on the hustings, enlightening marketers about the endless promise of CGA. It was in fact a month after the website's launch. Yet if you logged onto XLNTads.com, all you found by way of advertising were samples, in a half dozen product categories, of non-existent brands: ChocoBana candy bars, XCell mobile phones and so on. This was because, at the time, no actual brands had signed on.

"That's the killer right there," Perry said. "Until we have someone up on the site, you struggle from a sales standpoint." It's a typical chicken-and-egg situation, in which interested prospects balked because no other interested prospects had acted. Perry described the sort of exchange that plays out during sales calls:

Prospect: "What kind of ads can I expect?"

XLNT rep: "Uh … I don't know…. funny ones?"

At least by this point the website was up. The fake ads for the fake brands helped customers visualize what CGA might yield. "Up until then," Perry sighed, "we had nothing to show them."

What a difference a few really terrible fake commercials can make. After only four months, the cartons of copy paper by the front entrance had been replaced by two handsome leather chairs. Two previously vacant offices in the suite were buzzing with activity and Perry was promising that by fall the site would host at least two and as many as six honest-to-God brands. One was True.com, an online dating service. Another was Bomba, an Italian energy drink. Four other companies — mainly in the second tier of the Leading National Advertisers but all major brands — had contracts in various stages of the approval process.

"We're getting there," Perry said.

They were getting there with the continually evolving strategy, based on the advice of "creator" panelists like Kevin Nalty and the often ugly outcomes XLNTads has witnessed involving CGA pioneers in the real world. That includes Heinz, whose contest attracted the Ketchup Man/ Prince of Darkness entry and another featuring a guy shaving with ketchup.

"They went out to the world. They did *USA Today* ads. They put it on their packaging. As a result, they got well over a couple of thousands entries. And 50 – 60% of them were junk — but somebody had to look at every single one of them. We're trying to avoid that trap."

Trying to avoid — in their consumer-generated ads — those inconvenient, you know…. consumers.

XLNT also hired community coordinators to beat the bushes of the amateur video world in search of the 500 premium contributors so that brands using the site to host contests will not have a Heinz-like inundation of crap from the clueless, but rather a ready supply of dependably watchable material. And surely that would make CGA contests run a lot more smoothly with a lot more satisfactory results. But doesn't that sort of obliterate the whole point of CGA? If contests are dominated by a relative handful of ringers, doesn't that defeat the purpose of a consumer promotion — i.e., promoting widespread engagement with the brand? Furthermore, you can scarcely tap the wisdom of the crowd if your first move is shrinking the crowd. If the concept is to widen and deepen the pool of idea generators, cultivating a Dilettante Elite would seem antithetical to the goal. Right, Neil?

"Not so, dear friend."

Perry did the math a slightly different way. If, at any given time, XLNTads.com is hosting 10–15 CGA contests, the anointed 500 will be spread more or less evenly among them. Their presence, he said, will merely guarantee a sort of content liquidity. They exist for the same reason beggars and buskers toss their own coins into the hat the moment they open for business. Passersby, believing others have preceded them, are more likely to toss in money.

But it's not just the dynamics of video-hobbyist behavior he was trying to capture. It was also the medium's essential authenticity.

"CGA to me is also about the advertising style. It's about handheld camera, low-cost/lowdown production, simple ideas and simple executions. That's what makes it so relevant and that's what makes it feel so different from the Madison Avenue way of doing things."

So, wait, Neil … are you sure you mean "authenticity" and not "amateurishness?"

"That's my story," he said, "and I'm sticking to it."

But Not Indefinitely

Flash forward almost two years. A consumer generated commercial — again for Doritos — has won the *USA Today* Super Bowl Ad Meter contest. Dave and Joe Herbert, two unemployed filmmakers, won $1 million for topping the pros. This was no mean feat. To do so, the brothers had to dethrone Anheuser-Busch, whose Super Bowl ad for years had dominated the Ad Meter through subtle, sophisticated humor — such as sex jokes and exploding horse flatulence. But then came the Doritos CGA, in which a guy uses a "magic" snow globe to smash the glass of a vending machine, and later wings it right at a blustery boss's nuts, teaching us:

1) that consumer-generated advertising had come of age
2) that Americans simply can't get enough of guys taking shots to the crotch
3) even more than horse farts.

This historic triumph would seem, on the face of it, to be fabulous news for XLNTads: genuine and high-profile validation of their founding premise, albeit resulting from a promotion they had nothing to do with. Four months later, however, they had seen no particular improvement in the environment. Risk aversion was their fundamental obstacle on day one, and so it remained.

"Certainly the issue that we talked about way back then is still the issue *du jour*," Neil Perry tells me in May of 2009. "There continues to be a great reluctance to jump into this by the major brands." Not to mention their agencies, who Perry believes could benefit from XLNT's services for small projects too unprofitable for shops to produce, but who Perry believes feel threatened by CGA.

That said, in the intervening time, much also had changed. For one thing, XLNTads is no longer headquartered in post-industrial Conshohocken. It now operates out of Wynnewood, a leafy, affluent Philadelphia suburb that has been home over the years to Walter Annenberg, Kobe Bryant and my mom. The new digs are situated in an 80 year-old frame house, lodged between a bank branch and a video store, a stone's throw from the Paoli Local rail service — also known, famously, as the Main Line. No sign of Grace Kelly or Katharine Hepburn, but Conshohocken it ain't.

Another big difference was the accumulated portfolio. Faintheartedness notwithstanding, several top marketers had indeed dipped their toes in the water. "We're doing one now for SmartScoop, an automatic cat-litter box for Our Pets," Perry says. "We just completed a second program for Nestle and we're just about to start a third. We started with the 100 Grand candy bar. Then we did Stouffers Corner Bistro Stromboli and Flat Melts. (Get 'em in your freezer case; they're delicious!)

"For Anheuser-Busch, we have completed four programs. One for Bud Light. We followed it up with one for Natural Light. And we did two separate projects for their designated-driver program. And we're talking about one if not two more.

"The last one that's a repeat customer is Callaway golf. We just completed one for the Diablo driver. We're talking to them now about their second program."

The resulting ads themselves were not, from a creative and strategic point of view, necessarily great triumphs. The ads, for instance, have not been televised; they exist only online, and few have had any ongoing presence even there. Indeed, with the exception of one edgy and fetching Diablo spot for Callaway Golf, they are all ... (Let's see, how would one put this if one were not wishing to bruise the feelings of all involved?) ... remarkably unremarkable. When Perry refers to his creators as "the B team, producing B-level creative," grade inflation is occurring. Nonetheless, the spots were produced, bought and run as is. These are not inconsequential successes. When last I visited, for instance, the work on the website was mainly for brands that did not actually exist. Nor does the progress end there. In the fall of 2008, XLNTads launched a parallel website called Poptent.net, an online community for ad creators and the hub of the new business strategy. Instead of positioning consumer-generated ads as mere contest fodder, the founders have decided to remake the company as a clearinghouse—like crowdSPRING—for crowdsourced TV spots.

"We've totally walked away from the contest model," Perry says. We've also walked away from big terms such as 'use this to engage your consumer.' What we've evolved to is a semi-professional and professional network of low-cost, high-quality video creators. It's basically crowdsourcing at its best. We've been able to drive up our level of creators to 9200, and we have started specifically trying to upgrade the quality of our creators. So now we are looking at less of the amateurs and more of the pros and semi-pros. Now about 30% of our community is small boutique shops. These are 2- , 3- , 4-person groups that are trying to break in. They can send their reel to General Mills and get it back in a brown envelope and a letter saying 'We do not accept unsolicited material.' Or they can do a Poptent assignment and at least know some national brands are looking at their work."

Perry was delighted, for instance, when Callaway sought permission to contact the *auteurs* behind its excellent XLNTad ad directly. The company so appreciated the work, it wanted to cut a deal. This was swell for the creator and — though he's cut out of any piece of the action — swell by Perry, too.

"We get some huge benefit out of the communication of that fact to our creator community," he says, "and it will probably increase the number of creators we get 2-fold." Not an insane rationalization. For instance, the McNuggets rap, when it was a mere CGA prank, had about 50,000 YouTube views. By the time McDonald's was done plastering it all over the tri-state airwaves, the online hits had multiplied by a factor of 50. Aspiring comics, rappers and ad-makers need visibility in the way vampires need blood and apparently everybody else needs a ShamWow.

The most important new development, though, is the substantial automation of the website. The low cost of buying a TV spot is made even lower via XLNTDirect, which posts and tracks assignments much in the way crowdSPRING does for graphic design projects. The cost of participating is reduced from $25,000 (plus $7500 for each spot chosen) to $9500 (plus $7500 each). A marketer would upload the creative brief and required "assets" — such as logo, product stills, theme music — and the rest is left to the CGA community.

"We don't review the ads when they come in," Perry says. "We don't recommend winners. Our sole involvement is giving them access to the creators and we help them with the paperwork at the end. We even give them a format for what the creative brief should look like. Contracts and agreement letters are standardized. It's sweet and easy. We can get them an ad for $17,000."

He still thinks this will be inviting for large marketers that also have a portfolio of small brands, as opposed to the long tail of small businesses who suddenly could actually have their own TV commercials. Still, at the moment, the question isn't which segment will be his biggest customer. The question is: will there be *any* customers?

"We think potentially that in 12 to 15 months we could be doing 50 assignments a month."

Up from the current two. Or sometimes three. The company, he says, "is close to being in the black, flirting with it" but much in need of $3.5 million in new capital to introduce the self-serve program to those marketers most in need of it. He's talking about an ad campaign, perhaps even in my publication, *Advertising Age.*

"We built this," Neil Perry sighs, "but that old adage about 'they will come' is not necessarily true. We've got to get out there and do some marketing."

Whether he'd hire an agency to do that marketing, or post his campaign brief on Poptent, he does not say.

Chapter 10

THE POWERS THAT BE 2.0

S OMEWHERE A GURU IS SCOFFING. Or eye-rolling. Or generally feeling that the, duh, *very heart* of Web 2.0 has thus far here been given short shrift. After all, at this stage of the proceedings there has been page after page on widgets, consumer-generated ads, collaborative filtering, customer-relations management and word-of-mouth to a fare-thee-well, yet not one single chapter of this book explicitly devoted to social networks. How can I be so careless? What could possibly justify such a glaring omission?

Hey, good question!

Almost!

The fact is, there is no need to devote a chapter to social media, because every single thing in the art and science of Listenomics *assumes* social media. Facebook, MySpace, Digg, Twitter, Bebo, LinkedIn, FarmersOnly, HamsterSter (!), SocialAnxietyFriends (!!!), you name it. The digital connectivity and power redistribution inherent in social networks are indeed at the heart of the New World Order. To devote a chapter to them would be approximately like a history of the newspaper business devoting a chapter to ink. Dude, it's a given.

Furthermore, it is a given not just for, say, Procter & Gamble and Pepsi-Cola. It's easy to survey the digital landscape as we've been doing for 227 pages and imagine all the chaos as some sort of arcane media/ marketing issue. But that is to tragically miss the point. The point is that *everything* is turned upside down. For so long, large institutions dictated and the crowd — the consumer, the viewer, the reader, the voter, the taxpayer, the spectator — pretty much had to be satisfied with what The Man chose to dish out. Now redistribution of power is occurring in every corner of society, every darkened recess of the economy, where the principles of Listenomics apply across the board.

Governance. Education. Art. Intelligence. Scholarship. Medicine. Science. Religion. Journalism. Politics. In each of these varied institutional pillars of society, the hitherto Powers That Be are already becoming the Powers That Were as the organizing principles of the analog world increasingly lose their dominance — and relevance — in the digital one. I'll single out a few of those pillars here, concluding with the most disruptive manifestation of all: the digital revolution's effect on revolution itself.

Regrettably, it is these subjects to which I will indeed give short shrift — not because they lack importance, but because each is more than a book subject itself. Also, paradoxically, I'm loathe to try to belabor revolutionary changes that have been so widely chronicled and elsewhere discussed. For instance: politics. It's more than obvious that MyBarackObama.com was the apotheosis of social networking's intersection with election campaigns. It is the model for all candidates of every ideology at every level of government. As Bush advisor Mark McKinnon told *The New York Times*, 2008 was "the year campaigns leveraged the internet in ways never imagined. The year we went to warp speed. The year the paradigm got turned upside down and truly became bottom up instead of top down."

But you already know that. You know about the $500 million MyBarackObama.com helped raise. You know about the 13 million email addresses it gathered. You know about the million volunteers it mobilized. You know about the 200,000 events around the country it coordinated. You know about the tools it provided. You know about the movement it unleashed. So instead I'll review three watershed moments that *led up to* Obama's phenomenal exercise in community building—moments that forever have changed the relationship among the electorate and their would-be representatives.

The Vast Left-Wing Conspiracy

Remember, back in January of 1998, during the Bill Clinton impeachment debacle, First Lady Hillary Clinton going onto *Today* with Matt Lauer and complaining about the "vast right-wing conspiracy that has been conspiring against my husband" at that point for seven years? She wasn't being paranoid. As testimony of some participants later bore out—and the lockstep of Fox News Channel, Rush Limbaugh and the GOP attack machine made quite evident—President Clinton was indeed bedeviled by an organized partisan campaign of jaw-dropping scope and ruthlessness. It was very much a product of old media, like talk radio, and of politics-as-usual, like whispered smear-mongering about the supposedly suspicious suicide of White House aide Vince Foster. So here's what the Democrats did:

1) Whine.
2) Lose both houses of Congress and the presidency.
3) Eventually set about building a vast left-wing conspiracy. And an impressive one at that.

What the political left took advantage of was the convergence of growing discontent and a set of new tools: digital media. Partly by happenstance and partly by design, forces from various spectrum points of the political left separately cultivated online constituencies and coalesced to benefit the Democratic Party just in time for the 2008 election. The triumph of Barack Obama, though, was merely the culmination of a community-building, and fund-raising, exercise unprecedented in political history. And it began in 1998, about nine months after Hillary Clinton's bitter appearance on *Today*. It was in the autumn of that year that MoveOn.org was formed.

In the ensuing decade, MoveOn has become the prime engine of progressive organizing. At the time, though, it represented a single gesture by one center-left married couple, Joan Blades and Wes Boyd, who were sick of the impeachment circus and how it distracted the White House and the Congress from conducting the rest of the people's business. The single point of their online petition: Congress "should censure President Clinton and move on." This fairly unradical proposition quickly generated 100,000 signatures, emboldening the organizers to take the next step with their suddenly forming community: turning the grassroots into political activism. Fellow travelers were urged to contact Republican members of Congress and right-wing talk radio shows (often posing as Republicans), threatening to withhold votes from the GOP forever if it succeeded in booting Clinton from office. In the end, Clinton was acquitted in the Senate, though only five of 55 Republican senators voted not guilty on both counts against him. No Democrats crossed party lines, as would have been necessary to obtain a 2/3 majority.

The immediate political toll was hard to gauge. Clinton survived and retained high approval ratings in polls. Nine of the 13 leaders of the House prosecution retired or lost election over the next four years. Vice President Al Gore, though (or because) he put distance between himself and Clinton in the 2000 presidential campaign, lost (or didn't) to George W. Bush in 2000, who for four years enjoyed majorities in both houses of Congress. The only certain winner in all of this was MoveOn. org, which had accumulated tactical experience and a priceless email list. These it used to support Democrats in key districts, a strategy that in the 2004 election was finally availing.

By then, it had seized on another issue: the war in Iraq. Before the invasion in 2003, MoveOn.org once again floated a petition embodying a relatively modest demand: "Let the inspections [in search of weapons of mass destruction] work." As *The Nation* magazine's Christopher Hayes told me, following his 10-year retrospective of MoveOn, "They weren't saying, 'No blood for oil.' They weren't saying 'Renounce imperial conquest.' They were saying, 'Let the inspections work.' It was this very kind of middle-road path that they took."

Within a few years, as the war turned more calamitous, the middle road more or less disappeared — in American political discourse in general, and in MoveOn.org's politics in particular. The erstwhile voice of moderation, like much of America, had become radicalized by the missing WMDs, a "mission accomplished" false alarm, rampant anti-Americanism, the loss of blood and treasure, intractable sectarian violence, waste and corruption, Guantanamo, Abu Ghraib and a poverty of hope for a satisfactory resolution . Along the way, MoveOn's email list, its fund-raising capacity and its political clout only grew and grew. Though it has been the author of some obnoxious blunders — most infamously, rhyming Gen. David Petraeus with "betray us" in a newspaper ad that backfired in the fall of 2007 — it has also amassed 4.2 million members and become a force to be reckoned with.

In combination with such new-media outlets as TheDailyKos.com and *The Huffington Post* (not to mention the whole of MSNBC), MoveOn.org can share the credit with galvanizing a political left that had been fractured and impotent since the Lyndon Johnson administration. Meanwhile, another "conspirator," a Washington think tank called the Center for American Progress, was chartered explicitly to counter the impact of such conservative stalwarts as the Heritage Foundation, the Competitive Enterprise Institute, The Cato Institute and the Hoover Institution. Its offshoot, Media Matters for America, was created in 2004 to debunk the lies, propaganda (and inconvenient truths) bandying about the echo chamber of right-wing media. While the political right was marching — and defaming — in lockstep, says Media Matters' Eric Boehlert, "on the liberal side and the progressive side, there wasn't really anything." That was then. This is now.

If you wish to be awed by the intersection of political organization and online tools, visit MediaMatters.org, check out its latest content and then do a Google blog search to see how far and how fast it has been dispersed in the left-o-sphere. At this writing, I've copied and Googled a Media

Matters headline fragment "DHS Freak-Out" — about GOP reaction to a Homeland Security report on returning veterans — and found it repeated verbatim more than 200 times. This was in the space of 3½ hours.

Before the Scream

Howard Dean, before he was undone by the old media, was elevated by the new.

In 2003, partly on the strength of his unequivocal criticism of the war in Iraq, the then-governor of Vermont rose from obscurity to prominence among 12 early candidates for the Democratic presidential nomination. What got the national press interested enough to boost him in their coverage, however, was not his policy positions. It was his cash position. For instance, in the third quarter of 2003 — three months before the primary season — he raised just under $15 million. This shattered a Democratic record set by Bill Clinton in 1996, when in that period he raised just over $10 million. More to the point, Dean's war chest wasn't stuffed with big checks from fat cats, obtained by well-connected "bundlers" with their eyes on ambassadorships and cabinet posts. It was overflowing with $25 checks solicited online. And it was supercharged by "meetups" organized online. The "netroots" phenomenon — a term coined by blogger Jerome D. Armstrong — was born.

Pundits marveled at the campaign's ability, with a few web posts, to rally 3,200 supporters in Austin, Texas (i.e., Greater Crawford) in a matter of a few days. But the larger story was the ongoing virtual rally on the Dean2004 site, which was a proto-community — perhaps not as endowed with Web 2.0 functionality as MyBarackObama.com would be four years later, but certainly a meeting place and staging ground for the faithful. Here's what Doc Searles — co-author of *The Cluetrain Manifesto* — had to say about Dean's site at the time:

> *What makes a good poliblog? Human beings speaking in human voices ... Making the ends the means. Links galore. Cross-crediting where due. Permalinks. RSS. Involvement. All that stuff is good to see, regardless of the policies involved.*

By the time of the Iowa caucuses, the first major event of the primary season, Dean had catapulted to frontrunner status. Then two very bad things happened to him. First, he discovered the neither cash on hand nor polls nor the blessing of the punditocracy constitute actual votes. He polled third to John Kerry and John Edwards in the caucuses. Then, addressing his volunteers in the wee hours, and attempting to boost their morale for the upcoming New Hampshire primary, he let out with a sort of a shriek. It was the word "Yeah!" — as you might imagined it being shouted by Crazy Horse, on his way to slaughter the cavalry.

Now, I happened to be watching the episode live on CNN, and in context of the situation, and the crowd, and the near hysterical energy pulsing through the room, it didn't seem all that odd. But by the time cable news isolated it, and played it without context about 12 million times along with pundit speculation about the candidate's sense of dignity and emotional stability, all the governor's meet-ups and all of his cash couldn't stop the campaign from being smashed to bits. A month later he dropped out of the race. But netroots campaigning went on to reinvent American politics — not top to bottom, but bottom to top.

Tour de Farce

I fly to and from Washington, D.C. a lot, so I run into people. In the fall of 2008, just before the presidential election, one of those people was Sen. John Kerry, who four short years earlier had lost to President Bush in a race that should have put the unpopular president at risk against a used-car dealer, a raving Trotskyite and/or a corpse.

There are many explanations for Kerry's loss: his aloofness, his wealth, his un-demure wife, the grotesque smear campaign waged against him by the Swiftboat Veterans for Truth, the electorate's historical reluctance to dislodge a commander in chief in wartime. But as good an explanation as any is Kerry's own two left feet: his uncanny knack for saying and doing just the wrong thing at just the wrong moment. Such as challenging the notion of rich-guy detachment by being photographed windsurfing off Nantucket, or explaining his change of heart on Iraq-war funded by saying "I actually did vote for the $87 billion before I voted against it." Or by opening his official candidacy at the Democratic convention by pulling his naval salute out of mothballs and declaring, like some sad has-been at the VFW bar, "I'm John Kerry and I'm reporting for duty."

As Democrats everywhere cringed.

So aboard the Delta Shuttle, I asked Kerry what I've wanted to ask a presidential candidate for years: "What is it like to have to measure every word that comes out of your mouth, knowing that any mispronunciation, any fact error, any ill-considered joke, any burst of impolitic candor could come back to haunt you? Isn't it torture to be so perpetually on guard?" This verged, I grant, you, on being an unfair question, along the lines of "Do you still beat your wife?" If he were to answer me frankly and forthrightly ("Bob, it's a nightmare. I can't tell a dirty joke. Ever. I can't laugh at one. I also can never react in anger. If I smash my thumb with a hammer, I have to focus-group my response. I can never contradict myself. Or even change my mind. It's a deal with the devil; I get six years in office in exchange for my freedom of speech. There should be a politicians' Miranda warning: 'Anything you say can and will be used against you in the court of public opinion.' Frankly, it sucks."), then I as a journalist would report it and he'd spend the rest of his career trying to explain why he so resents his lofty elected position and the hardworking voters who offered him their trust, blah, blah, blah. No wonder Kerry seemed so annoyed as he dismissed the question.

"It's just a matter of discipline," he said.

What happened next, I'm embarrassed to say, was me breaking out in a shit-eating grin. It was meant to seem affable, but it was obviously triumphal. "Well, haven't you just agreed with me?" I said. "I'm asking you what all that discipline does to a man." At this point, I witnessed something painful and hilarious: a politician struggling to form a response that would not concede any part of my point — for that would implicitly be an admission of disingenuousness — and to do it quickly, lest the very process of being careful in choosing his words confirm the premise he had just denied. Perhaps he was also recalling his infamous "botched joke" in California in 2006, when his discipline failed him, and his admonishment to students to do well in school lest they "get stuck in Iraq," was taken (or just spun) as a slur on the intellects of our brave men and women in uniform. The senatorial hard drive whirred and whirred. Finally he replied, mainly by changing the subject: "I think you can speak frankly and forthrightly on tough issues and I believe I have." Then he made eye contact with the flight attendant, who rushed to his rescue.

All of which I mention only to observe this: John Kerry is quite bright, and was campaigning at the time for nothing. What if you're on the stump, aspiring to high office, and you happen to be a moron?

Ladies and gentlemen, I give you George Allen.

In 2006, Allen was an incumbent Virginia senator running for a second six-year term against conservative Democrat Jim Webb. It was a crucial race for several reasons. For the Dems, a Webb victory offered a glimmer of hope for a Senate majority. For Allen, the race was a dry run for 2008. The state of the GOP at that time — as now — was so bereft and chaotic that Allen was regarded as a frontrunner for the Republican presidential nomination. It seems unimaginable now, for the man is a genuine nincompoop. (Think: Gomer Pyle, minus the gravitas.) Anyway, on August 11, Allen was at a campaign rally in the southwestern Virginia town of Breaks. This was a stop on what he called a "listening tour" — which is supposed to be about hearing the concerns of the ordinary voter, but in Allen's case was just an opportunity to badmouth Webb and lump him in (wildly incorrectly) with the liberal elite. Here is what he said:

"My friends, we're going to run this campaign on positive, constructive ideas. And it's important that we motivate and inspire people for something."

We know these were his exact words, because the whole stump speech is on tape. But the tape was not shot by a journalist (such as the one on hand in November 1988, when the as yet undisciplined Sen. John Kerry hilariously remarked, "Somebody told me the other day that the Secret Service has orders that if George [H.W.] Bush is shot, they're to shoot Quayle.... There isn't any press here, is there?") In Allen's case, the video recording was done by a volunteer for the Webb campaign. For the purposes of what they call "opposition research," a University of Virginia student named Shekar Sidarth was capturing the whole spiel, little knowing that the next words out of Allen's mouth would be about him.

"This fellow here in the yellow shirt, Macaca, or whatever his name is, he's with my opponent. He's following us around everywhere. And it's just great! We're going to places all over Virginia, and he's having it all on film and it's great to have you here. You show it to your opponent [sic] because he's never been there and will probably never come, so welcome.... let's give a welcome to Macaca here. Welcome to America and welcome to the real world of Virginia."

A few pertinent facts: 1) Sidarth, south Asian by extraction and swarthy of complexion, was born and raised in Virginia. 2) *macaca*, a reference to a jungle monkey, is a racial epithet along the lines of "wog," especially in North Africa, where Allen's mom happened to hail from. 3) George Allen, a popular incumbent, lost the race, and the Senate, for the GOP and any chance to credibly run for president. This not just because the Webb campaign quickly notified the *Washington Post*, which it naturally did, but because 700,000 people subsequently viewed the episode on YouTube. As BuzzMachine's Jeff Jarvis later summed up the situation: "Candidates, be warned: You will choke on your forked tongues."

Or, put another way: John Kerry was wrong. No politician dares utter an unmeasured word, much less say something disqualifying. Because campaigns no longer need fear that some stray journalist will be around to record the gaffe; they must reckon with the absolute certainty that a volunteer with a camcorder, or a crowd member with a cell phone, will capture the moment and distribute it to the entire world. In a connected world, everyone at every rally is a journalist (recall when Obama was caught in a truth, about white working-class voters who "cling to guns and religion"), and nothing will ever be the same.

And Not for Journalism, Either

As the first two chapters of this book made frightfully clear, it's all but over for the newspaper business, offline and on. There is no business model currently existing to underwrite the editorial headcount required to produce daily journalism of the sort we've grown accustomed to for the past few centuries. That's the bad news. The good news is that, thanks to the very digital revolution that's killing the news business, there's another hive of journalistic worker bees to take advantage of — namely: everybody. From bloggers to spot-news eyewitnesses to online readers deputized to do basic reporting, the distributed world offers human resources such as no publisher has ever had on the payroll or ever could. These folks may not be trained at Columbia or Missouri or Medill, but they offer the advantage of unlimited aggregated man-hours at zero dollars per hour, plus ubiquity. The BBC, *The New York Times* and CNN among them have a lot of reporters and stringers — thousands — but thousands is fewer than hundreds of millions. All future models of journalism in various ways rely on this Infinite Newsroom. But this is not mere theory. Online and off, they — which is to say "we" — have long since been mobilized.

Most famously, the earliest news reports and the earliest pictures from the December 2004 Indian Ocean tsunami, the July 2005 London subway bombings and the December 2008 terror siege in Mumbai, India were dispatched to a horrified outside world via text messages and camera-phone photos from ordinary citizens at the scene. Similarly, the unprovoked shooting of an unarmed, handcuffed man by transit police on an Oakland, California subway platform was videoed by bystanders and uploaded to YouTube — whereupon it finally got the attention of local media and pressured the government to open an investigation. What these episodes make clear is that, from now on, we are all of us embedded reporters — embedded not by the Pentagon into some Basra armored battalion but by Best Buy into our daily lives. It just so happens that nearly all of us carry tools to document news in progress. That's how, in the aftermath of the London bombings, for instance, the BBC performed well, yet Wikipedia performed better, gathering and synthesizing up-to-the-moment information from both on-the-spot witnesses but the whole world of informed sources.

A similar phenomenon had taken place during the tsunami seven months earlier — along with a spontaneously forming disaster-aid infrastructure built within hours along the Howard Dean model. I spoke at the time to internet guru and investor Esther Dyson, who immediately understood how technology invested individuals with power previously enjoyed only by mammoth institutions "to broadcast, to find things, to do research, to get heard … across the entire internet. And when something huge like this happens, that just becomes much more apparent." Because online connectivity — unlike, say, a TV network — can scale, exponentially, almost instantaneously. Still, catastrophe-in-progress is only one relatively narrow element of the journalistic portfolio. There is, after all, so much more that we godless media parasites do to maintain our tenuous attachment to the body politic. One is basic reporting of the workings of government at all levels. Another is analysis and opinion. Another is service journalism, also known as "news you can use." And then there is the category we call "enterprise," comprising everything that we on our own volition just start wondering about.

Such as, "Hmm, is there asbestos in the walls of our elementary schools?" "Why did the new interstate interchange get re-routed to where the governor's brother-in-law owns 400 acres and the biggest motel?" "Hey, the chamber of commerce is always bragging about our symphony orchestra. How many different people have actually attended a concert in the past 10 years?" "As mortgage rates go lower, are the re-financings benefiting people who had been sucked into onerous sub-prime terms, or well-heeled folks further feathering their nests?"

For obvious reasons, a lot of enterprise reporting tends to fall into the sub-category of investigative. What is less than obvious is that journalistic investigations don't generally involve clandestine meetings in parking garages with mysterious Deep Throats or high-level leaks about black-site CIA prisons in Eastern Europe; they more typically involve sitting down with huge reams of public records and trying to sift out the truth obscured within. This exercise has all the glamour of term paper-writing. It is also so time-consuming that many a worthy investigation goes unconducted, or prematurely aborted, because of limited reportorial resources. As traditional news organizations spiral ever closer to oblivion, that problem has gotten dramatically worse. It's hard to justify assigning your best reporter to an open-ended stay awash in courthouse records when you're not even regularly covering the county commissioners. This is where the Infinite Newsroom comes in very handy. The crowd has aggregated curiosity, IQ to spare and all the time in the world.

It also walks some roads Woodward and Bernstein seldom traveled. This doesn't get much publicity, but one of the earliest and purest examples of crowdsourced reporting began in 2001, when blogger Robert Niles of ThemeParkInsider.com started a feature called "Accident Watch," which is exactly as it sounds: reports about accidents at parks around the world. More than 100 have been filed and verified since Niles asked readers to report on incidents they'd witnessed or learned of. As of this writing, the most recent was this, from March 20, 2009 on "The Eighth Voyage of Sinbad" attraction at Universal's Islands of Adventure in Orlando, Florida.

> at the 3:30 show. The very begining of the adventures of sinbad show the actor which we assumed was sinbad started from a highwire slide from the stage to a platform and fell into the audience and partially on stage props located in the middle of the audience as well … the cord he swong down from snapped and fell at least 1 1/2 stories from mid air into the stands..once that had happened the audience was removed immediately and nothing was said afterwards.… went to check the net and newscast about it and nothing.. saw a new chopper show up for a few moments after the incident but not sure if it was pertaining to that accident or not.

Rather Embarrassing

Not exactly the scoop of the century, but it's worth noting that Accident Watch "swong" into action on this story before the *Orlando Sentinel* did. Probably the first big scoop attributed to the blogosphere was the debunking, in September 2004, of the central pieces of evidence in a CBS News *60 Minutes* story about President George W. Bush and his Vietnam-era service in the Texas Air National Guard. Based on memos obtained by his producers, Dan Rather concluded that the ultra-connected young George Bush had received preferential treatment and political protection to both get into and stay active in the Guard. Ever suspicious of the "liberal media," and especially Dan Rather, conservative bloggers got to work to defuse the bombshell.

Didn't take long. Within hours, the right-wing website FreeRepublic was questioning the authenticity of the purported smoking-gun memos. Quickly a blogger with the handle "buckhead" asserted that the typeface was anachronistically modern. By the next day, the blog Powerline narrowed the font down to the Microsoft Word's version of Times New Roman, complete with the itty-bitty superscript "th," as in "187th Tactical Reconnaissance Group." There was no such thing on a circa 1972 Smith-Corona, needless to say — although evidently not entirely needless to say, inasmuch as CBS producers failed to notice it. None of this itself disproved the premise of the *60 Minutes* report, but it certainly blew the supposed documentary evidence all to hell, along with the careers of four CBS producers and Rather himself. It helped not at all that former top CBS News exec Jonathan Klein had chosen to cast the dispute as a *The Mouse That Roared*-esque assault on the mighty Tiffany Network by a ragtag band of impertinent upstarts. "You couldn't have a starker contrast between the multiple layers of checks and balances [at CBS News] and a guy sitting in his living room in his pajamas writing," Klein declared.

Oh, couldn't you? As FreeRepublic's Kathleen Parker observed at the time, "The implication that bloggers are slacker dust bunnies has delighted bloggers, the best of whom are lawyers, professors, scientists, renegade journalists and techies of various sorts, such as the brothers Johnson (Charles and Michael) at LittleGreenFootballs[.com], whose years of experience in state-of-the-art graphics and Web design at the 'pixel level' enabled them to quickly duplicate the CBS memos and demonstrate their likely origin on a very modern computer." Yeah. What she said. Two months later, with the substance of the CBS story as yet unrebutted, President Bush won re-election.

I Shot the Sheriff, But ...

In that incident, the president was the beneficiary of blogger vigilance. Three years later, his administration would not be so fortunate. This was in March 2007, when Josh Marshall of the blog TalkingPoints-Memo wished to discover whether White House had strong-armed the Justice Department into firing politically wayward U.S. attorneys. Under pressure from Congress and the media, the administration had released more than 20,000 White House and Justice Department emails—a document dump that would choke any news organization or congressional committee staff obliged to sort through it. Enter Talking Points Memo, where staffer Paul Kiel directed readers to the email cache and told them to have at it, explaining: "Josh and I were just discussing how in the world we are ever going to make our way through 3,000 pages when it hit us: we don't have to. Our readers can help."

They did that. The TPM faithful were quickly able to find gaps in the record and holes in the administration's story, revelations that fed more Congressional hearings that in turn led to the "resignation" of Attorney General Alberto Gonzales. It was a breathtaking display of the crowd completing in days what even a team of staff reporters might not finish in months. It was also an extremely rare example of citizen journalism not tied to an existing traditional news organization. Although those are increasingly productive, too.

In the summer 2007, for example, residents of Cape Coral, Florida, got some unwelcome news. The assessments for a new water and sewer system were rising from $12,000 per home to, in some cases, $40,000.

The local *Fort Myers News-Press*, of course, jumped on the story. But the paper didn't just deploy a few reporters. Online and in the print editions, it also asked readers to get involved. Within hours, says executive editor Kate Marymount, information came pouring in, from assessment letters to blueprint analysis, to you name it.

"Sixty-five hundred different individuals in the community contributed information," Marymount told me. "On the second day of this project, we got an email from someone several states away who saw it on our website and said, did you know that an audit had been done and not released? Would you like a copy?"

A copy of a government audit, previously undisclosed, about out-of-control project costs? Oh … okay. The paper posted the audit, whereupon all hell broke loose, immediately — and pretty much ever since. It's an ongoing investigative story, a collaboration between the professionals at the paper and citizen deputies documenting failed oversight on a grand scale.

"We would have gotten pieces of it," Marymount said. "I don't think we would have gotten all of it. We can't be in every neighborhood. We can't have access to every document."

Therein, again, the irony. Though the internet has fragmented audiences and decimated advertiser reach, it has simultaneously afforded news organizations more journalistic reach than they ever had before. Here's my WNYC radio colleague Brian Lehrer in 2008 speaking to the audience of *The Brian Lehrer Show*, commencing a project to determine where New Yorkers are most likely to be price-gouged:

> *Price disparities of common food items in the New York area. The assignment is this. Go to your local grocery store, any local grocery store, and just check out the price of three goods — milk, lettuce and beer. We'll tell you exactly what kind in a minute. Once you have the prices, just go to our website and report your findings in our special comments page for that. Pretty simple, right?*

"That," says Jeff Howe, who coined the term, "is the future of crowd-sourced journalism" — because it doesn't ask anyone to commit journalism. It asks them to fan out to gather some basic facts. "Thy don't want to write a big feature on a change in a zoning law any more than they want to go back and rewrite the term paper they got a D on in college. But what they will do is they'll say, 'Huh, man, I want to know if I'm paying too much for a head of iceberg lettuce.'"

By the way, the answer is: not in Glen Rock, NJ (88 cents) but probably at the Food Emporium in Manhattan ($2.49) — a disparity due in some part to unscrupulousness and in some part, like all price gouging, to simple supply and demand. I raise that point only to observe, with all respect to Jeff Howe, deputizing listeners to gather data is only half the future of crowd-sourced journalism — the supply side, to be specific. The other half is the demand side, wherein the vast citizenry informs news organizations know what *it* wishes to be informed of, and about what it already knows. On the most basic level, this was illustrated by the April 2009 Digg interview with Trent Reznor of Nine Inch Nails, a Q&A in which the Qs were those suggested by Digg's audience and which had been propelled by user voting to the top of the site. (Hilariously enough, here was the top question: "Your business model still primarily involves selling music either digitally or physically. Why haven't you embraced advertising as a business model?")

The demand side is served far more broadly in an initiative, pioneered by American Public Media and its Minnesota Public Radio (and also embraced by WNYC, among other news organizations in eight states) called Public Insight Journalism. In this model, sources of news tips, story ideas and specific expertise are cultivated in advance and drawn upon during the ordinary course of business. This offers the advantage not merely of expanding the newsroom Rolodex far beyond the usual suspects, it opens up the editorial process to the audience — so that the day's report reflects not just what the beat reporters flushed out and what the editor thought of while brushing her teeth, but the observations and concerns of the actual community.

"The belief is that the audience with first-hand experience and knowledge should be able to inform news coverage," says Chris Worthington, news director at MPR. "You just need a system to connect with your readers and audience."

The system is to collect volunteers — more than 30,000 so far in Minnesota and 90,000 around the country — willing to be occasionally queried about events large and small. They supply basic information about themselves, data which is expanded with each subsequent contact into a virtual dossier. For instance, when a single-engine plane flown by Yankee's pitcher Cory Lidle struck an East Side building in October 2006, early speculation turned to possible design flaws in the plane — manufactured by Cirrus Aircraft in Duluth.

"We screened the database for plane pilots," Worthington recalls. "Then we screened it for people who have flown that plane. In, honest to God, an hour and a half, we were able to write two solid paragraphs that were able to knock down the speculation about the plane, sourced by eight or nine pilots who said it's a wonderful plane, it's airworthy, I've flown it for years."

The subsequent investigation by the National Transportation Safety Board indeed concluded that the fatal crash was due to pilot error. Still, Worthington prefers to focus not so much on the system's ability to react to news as to anticipate it — "to get ahead of the news, to get ahead of trends, to spot things that we're facing."

In one case, a routine query to its community about dealing with soaring gasoline prices yielded a story about a growing gap in home care for the disabled, because the providers could no longer afford to drive from client to client. Another query about an increase in the number of Minnesota schools failing to meet federal standards unearthed a glitch in the test-result reporting system. News reportage is, by definition, reactive. Public Insight Journalism is, by definition, proactive.

Please, though, mistake none of this for a sustainable business model. Despite some interesting academic experiments, such as the Assignment Zero project at New York University, nobody has figured out a way to conduct "citizen journalism" except on a small scale, and certainly not at a profit. As new-media consultant Mark Potts bluntly assesses economic reality: "The elephant in the room is there's not a business model. We're still sort of in the 'Let's put on a show!' stage of user-generated, hyper-local citizens' journalism. 'Hey, kids, let's put on a show! I've got Mickey! I've got Judy!' We've got to find a way to make a business out of it if you're going to sustain it. Otherwise, it's a hobby."

Andy Hardy Prevents Malaria

Potts isn't necessarily wrong, but perhaps he is missing something—like a moment in human history in which progress becomes less dependent on the bottom line. Not everyone and everything turns on the profit motive. If we know anything about the distributed world, it is that it breaks down distinctions between vocation and avocation, between professional and dilettante, between the monetizable and the commonweal. Just for instance, a trip to west Philadelphia yields a hidden-in-plain-view insight into the art and science of Listenomics—namely: the science part. No exotic venue this time around. The scene is an urban college campus, Drexel University, where associate professor of chemistry Jean-Claude Bradley, shows off his laboratory. I am, frankly, stunned. If the last time you were in a chemistry lab was in high school, you'd be surprised to set foot in a modern one situated in a significant research institution.

It's exactly the same.

Same vaguely sulfury odor. Same linoleum-tile floor. Same banks of louvered fluorescent fixtures overhead. Same rows of blacktop lab benches, each fitted with gas jet for the Bunsen burner. Same jars of commodity chemicals. If it weren't for some pricey gizmos—the Rotavap, the Branson 1510 Sonicator, the Vortex-Genie 2 mixer—and the hum of the industrial-strength fume hood, you might think Mrs. Greenspan was going to walk in any second to give a pop quiz on benzene.

One major difference, though. Here Mrs. Greenspan would be flunking people left and right, because—in this lab—everybody looks at everybody else's work.

"What we do is: the entire process is wide open" says Bradley, a boyish-looking, mop-topped, somewhat Mickey Rooney-ish 40-year-old in a t-shirt and jeans, and one of the pioneers of the Open Science movement. In his lab, and among his network of fellow Moleculeteers, it's all for one and one for all.

Science, historically, has been a secretive pursuit. Yes, discovery builds upon discovery, but — until research is actually published — the incremental work of experimentation and observation, trial and error, hypothesis and result tends to be performed very much in the stealth mode, opaque to the outside world until such time as it is presented at a conference or in a peer-reviewed journal. And God help the poor sap whose years of research are scooped by a colleague, perhaps one halfway around the world, working on a similar or identical track with or without knowledge of the competition. Because "first" is first and second is often too late.

Science puts a great premium on individual accomplishment for all of the obvious career and ego reasons and perhaps some less than obvious — such as the ownership of patents on laboratory discoveries.

Even the crowdsourced science pioneered by InnoCentive, as discussed back in Chapter 8, is conducted under ironclad non-disclosure agreements, with resulting innovations owned by the sponsor with no obligation to publish for the outside world. But what if the competition model were replaced with, or at least augmented by, a cooperation model? What if science were not proprietary? Open-source software such as Linux proves that many heads can be better and faster than one. The only thing lost through collaboration is clear-cut ownership — which Bradley naturally understands is no small matter. Money and professional standing, even fame, are no less motivating to scientists than to the general population. And Nobel Prizes aren't awarded to collaboration networks. But those things don't matter equally to everybody, and they don't necessarily serve the goal of efficiency.

"It's a values question," Bradley says. "There is a core group of people who believe that the whole scooping problem is not as large as the opportunities we are missing. We're not interested in intellectual property."

What they're interested in is intellectual aggregation. Hence Bradley's project: The Open Notebook Science Challenge. It's a contest, like the Netflix Challenge, minus the $1 million jackpot. Instead it awards $500 at a time for contributions — from within an ad hoc community — adjudged quality science on the path to greater discovery. The ultimate goal is to produce cheaper anti-malarial compounds, but some precursor research involves the basic — and characteristically tedious — science of measuring the solubility of various compounds (say, nitrobenzaldehyde) in various solvents (such as methanol or acetonitrile). The process involves using magnetic resonance to quantify dissolution at increments of temperature.

This could be done by Bradley himself, along with his students, or the labor can be spread out among various of Bradley's academic collaborators — in this case chemistry professors and their students at Oral Roberts University, Indiana University and England's Southampton University. As we speak, he opens his laptop to a wiki page, where he consults the latest entry by a student named Matt.

"Eight minutes ago," he says, "Matt made a change to experiment number 77. Hmm ... what did he do?"

What Matt had done was put a sample of diphenylacetic acid in methanol, dissolve it and subject a quantity of the solution to a magnetic field. Then he counted molecules to formulate a solubility value. Or something like that. Technically, Bradley had more or less lost me after, "Nice to meet you." The point is, he was able to keep abreast of Matt's progress and to comment — as he can do for every event logged by every collaborator, on basic research or ambitious experimentation, because every step is recorded online in an open forum. And not only the Open Notebook wiki forum. The solubility values immediately are posted on the Wikipedia entries detailing the chemical properties of various compounds — available for the world of researchers to see. Whether cheaper anti-malarials are forthcoming or not.

"That's what motivates me," Bradley explains. "People use it."

Dear God, Now What?

Oh, my but the things we can now do online: Record the solubility of diphenylacetic in methanol. Pore over emails to make a liar out of the attorney general. Play Halo. Lose the kids' college fund at Party-Poker.com. Unload that ugly sofa on Craigslist. Find sexy singles in your area. Or, laugh until you wet yourself reading Failblog.org. (I won't even describe it. Just go there. You will not be disappointed.)

Or, if the time of year is right, there's always a nice Passover cyber seder, in which you can join real Jews at a virtual table and experience the bittersweet remembrance of Exodus without actually having to eat gefilte fish. Or, if Judaism isn't your cup of theology, perhaps your avatar is ready to undertake a virtual hajj, on Second Life, where many thousands of Muslims have donned garments made of pixels and kissed the digitally rendered Black Stone at a graphic representation of Masjid al-Haram.

Or you can worship Jesus Christ — or study the bible, or volunteer, or conduct a faith mission — at LifeChurch.tv, where there is no parking-lot jam after the service. It takes place online.

"Religious people since the advent of the printing press have been adopting the latest technology," says Heidi Campbell, assistant professor of Media Studies at Texas A&M University and author of the forthcoming *When Religion Meets New Media*. "When I look at my students, their understanding of what it means to be in community, identity, ritual even religion is changing because of their relationship to technology. They have a different idea of how their social world can be constructed."

So they do not necessarily think it insane to propose that some functions of organized religion can be achieved, and in some cases improved upon, online. For starters, virtual congregations are not limited by geographical boundaries, which you pretty much can't say about the Five Corners Assemblies of God Church of Anytown, or whatever. They can also be organized by demographics and affinities — along social-networking's vertical lines — as opposed to the relatively random distribution implied by the limits of neighborhood and denomination. Moreover, you don't need a building. You don't need pews and stained glass. Or utilities. Or a kitchen. Or a bus. Or a janitor. Or even, necessarily, a pastor. All you need is an ad hoc community of like-minded (or not like-minded) people to gather, in person or online.

All of which, if you think about it for even an instant, is potentially as disruptive to the God industry as crowdSPRING (Chapter 8) is to graphic design.

"Everything you know about digital age about advertising and media and journalism, pretty much fill in the blanks for religion," says Andrea Useem, author of the 30,000-word white paper "The Networked Congregation."

"There's a decoupling of functions: worship, socializing, community, education, something to do Sunday morning, and building up an identity for your children. If you wanted to do that [historically], you pretty much had to find a church, but now there're options."

One that especially intrigued Useem was Meetup.com, an online organizer of offline organizations across all areas of getting-togetherness, from arts to pets to hobbies to sports to parenting to spirituality. "The whole amazingness of Meetup.com is you can find likeminded people in your area and meet them face to face," she says. In her case, she met up with the Alexandria (Virginia) Gathering of the Beloved, a group of 83 Christians who meet weekly in members' homes, with a decidedly unrevolutionary agenda:

> We are a group of people who desire to know our Lord in a deeper way individually and corporately. When we gather together our purpose is for our Lord to be expressed as we seek to know Him as our Head. Our meetings are an open time of sharing, singing, praying, and fellowship bringing each person's portion of Christ to build one another up in love and unity in The Spirit. Christ is our unity, nothing more and nothing less. We do meet in homes, and often have meals together, but it is our desire to not just be a house church, or to try to do "church" a different way. Our desire is to be a locatable, tangible, expression of The Body of Christ.

Useem visited on a weekend evening and was blown away by the warmth, sincerity, ethnic diversity and generosity of the group. The service, she says, "was so alive, so heartfelt" — a quality she's found subordinated at her own place of worship to mundane practicalities. "Our mosque is always trying to raise money to pay off the parking lot, which is probably the least inspiring cause I've ever heard of." But if she hadn't found the Gathering of the Beloved so belovely, Useem had many other Washington, DC-area meet-up options: fellow Muslims, Jews, Catholics, Buddhists, Methodists, The Montgomery County Pagan Meetup Group, Fun Loving Atheists of Maryland, the Washington Secular Humanists Meetup Group, Gaithersburg Area Search for the Paranormal, Tribe of Beltway Shamans, and the Maryland Jedi Order.

None of these groups in and of itself exactly threatens the foundations of, say, the Vatican, but their vast theographic diversity, online organizing ability and low barriers of entry portend some market erosion in the brick-and-mortar category. Their proliferation also raises some interesting questions about what constitutes a place of worship in the first place. How can online experiences, or meet-ups at unconsecrated venues, provide sacraments and clerical authority? If the sanctuary is a fertile medium for spiritual receptivity, what makes the spirit move via 17-inch monitor? When an online congregant is in mourning or ill, who comes by with a casserole?

Of course, the issue isn't really what Religion 2.0 can't duplicate; the issue is what it can add, and for increasing numbers of worshippers around the world, the answer is "plenty" — including, but not limited to, global reach, resources and, very significantly, convenience. On a recent spring afternoon, I attended. (Me to wife: "I'm going to church now." Wife to me: "Dressed like *that*?") In my grubby shorts and T-shirt, I logged in just in time to hear the opening "hymn," a Pete Sanchez Jr. rock spiritual titled "I Exalt Thee," presented via three-camera video and performed by Chris Quilala.

> *For Thou, O Lord, art high above all the Earth*
> *Thou art exalted far above all gods*
> *For Thou, O Lord, art high above all the Earth*
> *Thou art exalted far above all gods*

And I exalt Thee, I exalt Thee
I exalt Thee, O Lord
I exalt Thee, I exalt Thee
I exalt Thee, O Lord

You have the benefit of the written lyrics. The musical effect, believe me, is such that you could be listening to the rock ballad and Quilala's kick-ass lead guitar for some good while before it ever occurred to you that religion was taking place. (As to whether this is good or bad for Christianity, I have no opinion.) When he is finished, any question of whether this is church immediately vanishes. It's "offering time" — time to tithe online, just like at MyBarackObama.com! — and then comes LifeChurch.tv founder Craig Groeschel. He's a nice-looking 41-year-old Oklahoman in faded jeans and untucked dress shirt, preaching to 140 countries at once about the Book of John, Chapter 9 — "a story that may help us to see God when you don't really understand."

Pastor Groeschel preaches, like so many clergymen before him, about strains on faith when calamity strikes. War. Natural disaster. Crime. ("How can a father get mad at his wife and shoot his own children in their sleep? Where is God in that?") As he speaks, pop-ups on the screen offer specific scriptural references, links for drilling down into the core theology and other readings on the subject. Meanwhile, on the right of the video window is a chat box, dubbed "mix and mingle," wherein congregants can discuss Groeschel's preaching or just hang out together. I asked them why they were logged on to LifeChurch.tv:

ChrisByers "to be part of a global church experience and to help others find Jesus."

Jeff Moore "wasn't able to go to church this weekend, but I wanted to 'go to church"

Triplatte: for those of us who are nowhere near a church. This is a good place to be."

As I was visiting with these folks, another 3100 worshippers around the world were online with us, too — a segment of the approximately

16,000 who click through during the 20 online services each week. This is fewer than the 27,000 LifeChurch.tv commands at its main location in Edmond, Oklahoma and its 12 satellites in six states. But, says pastor/innovation leader Bobby Gruenewald, it is a fraction of the "hundreds of thousands or even million people we hope to reach with the Gospel." He bristles at the idea that 3100 assembled e-congregants already constitutes a megachurch. "In the context of 6.7 billion people?" No, he says, LifeChurch.tv is "a microchurch with a megavision."

Nor is LifeChurch.tv alone. A South Carolina congregation called NewSpring is one of many using the web, in the words of its slogan: "To make Jesus famous, one life at a time." As web Pastor Nick Charalambous observed in an April 2009 blog post, "I think our current hang-up over whether you can have true community without physical presence is a colossal distraction from the reasons why. We've got to be where everyone is, use the communication tools everyone else uses, and share what Jesus has to offer wherever people expect to find 'knowledge for life.' And in our foreseeable future that will be defined by the web."

Not that any of this is an either/or situation. While clergy may worry about losing attendees — and revenue — to online entities with no parking-lot mortgages to pay off, Andrea Useem sees mainly opportunity for brick-and-mortar congregations to increase their reach, relevance and impact. Especially within mega-churches and other large congregations, social-networking tools can filter out discrete communities within communities: parents of Down Syndrome kids, say. Or the freshly divorced. Or those facing elder-care challenges. These members can be clustered for support or ministration, in turn fostering participation in congregational life as a whole. Useem points out that research, including Heidi Campbell's, suggests that those who conduct some aspects on their religious lives online tend to devote more — not less — time to their physical places of worship.

"If you're a pastor," she says, "that's a pretty big headline."

Certainly at LifeChurch.tv Pastor Gruenewald doesn't much sweat the question of whether online worship displaces or augments the traditional church; he sees it as a bit of both. But he fears not at all a loss of personal, flesh-and-blood contact in traditional pastoral relationships. For one thing, he notes, in today's brick-and-mortar megachurches, clerical connection is already often a one-way ticket. Secondly, LifeChurch.tv offers live, one-to-one prayer, online. Thirdly, LifeChurch.tv avails itself of opportunities to reach out to the spiritually hungry with uniquely digital tools. Like Google search. Yes, they buy Adwords that kick in when a service is up live. And the words aren't "worship," or "Christ" or "Gospel," either.

"We buy 'porn,' 'live sex,' 'triple x,' Gruenewald says, "to find someone at a time when they're feeling hollow. Just two weeks ago, someone wrote in and said they were searching for pornography and they came across this ad, and they said, 'I'm just so thankful. It was a weak moment in my life, and I thought I'd found what I was seeking. Instead, I found God.'"

There is Congregation and There is Congregation

So let's agree: social networks can be the medium through which people find faith. The opposite is also true. Lost faith can find its expression, and a very rapidly forming community, online. Zeal is not limited to worship, nor are the tools of organization granted solely to the pious.

In his seminal 2002 book *Smart Mobs*, Howard Rheingold observed that, for better or worse, "communication and computing technologies amplify human talents for cooperation. The impacts of smart mob technology already appear to be both beneficial and destructive, used by some of its earliest adopters to support democracy and by others to coordinate terrorist attacks."

He wrote, for example, of the 1999 World Trade Organization protests in Seattle, where anti-globalization forces used websites and cell phones to mobilize more than 40,000 demonstrators, some of them violent. He also looked at text message-fueled protests in Manila in 2001, where hundreds of thousands took to the streets for four days to demand — in the midst of a corruption scandal — the ouster of Philippine President Joseph Estrada. The so-called Second People Power Revolution ultimately forced Estrada from office. (Whether this was the ultimate expression of democracy over entrenched forces of corruption, or simply mob vigilantism, has never been definitively answered. Though in the late stages of the protests the Philippine Supreme Court legitimized the accession of vice president Gloria Macapagal-Arroyo to the presidency, ruling that "the welfare of the people is the supreme law," some have come to regret their participation in the spontaneous *coup d'etat.* Among them: the Philippine National Bishops Conference. This is owing in part to the lack of due process, and in part to the subsequent conduct of President Arroyo, whose authoritarianism and own corruption scandals have made her extremely unpopular throughout her rule.)

Since the publication of Rheingold's book, there have been many further examples of civil unrest triggered, or fed, by real-time social networking. The riots of disaffected, dispossessed Arab youths in the suburbs of Paris in 2005, according to the national police, were not utterly spontaneous, but coordinated through SMS messages, blogs and email — warning, for instance, where the cops were. The Zapatistas of Chiapas, Mexico — a substantially Marxist movement dedicated to the interests of indigenous peoples — have embraced a basic strategy of using new media to forge alliances with socialists, anti-globalists, human-rights advocates and anarchists throughout the world. Once armed revolutionaries, the Zapatistas have long since waged their "war" against the Mexican state entirely on the internet. In 2008, Taiwanese students protesting a new law curbing the right to assemble defiantly assembled — having mobilized via an online bulletin boards — and broadcast the protest live on the internet via their cell phones. Then, in April 2009, in Chisinau, Moldova, came yet another new wrinkle.

In protest of a suspicious parliamentary victory by the Communist party, thousands of demonstrators converged on government buildings. Confronted by riot police, the crowds turned violent and stormed the halls of power. Over the course of four days, rioters ransacked and burned parts of the parliament building and presidential headquarters. The blame — or credit — for the sudden mobilization was quickly given to new media, such as Facebook, LiveJournal and, especially, Twitter. In fact, on his blog "Net Effect," Open Society Institute fellow Evgeny Morozov declared the Chisinau uprising "The Twitter Revolution."

Some took issue with the coinage, on the grounds of provenance. Analysis of "tweets" on the subject, conducted by Ethan Zuckerman of Harvard University's Berkman Center for Internet and Society, suggested that Twitter itself had little to do with mobilizing the mob into action, or in coordinating the protest on the ground. For one thing, at the time of the protests, only a handful of Moldovans had Twitter accounts. Twitter did, however, help catalyze the reaction, and coalesce attention and support from around the world.

"Where I think they were enormously important is helping people, particularly people in the Moldovan Diaspora, keep up with the events in real time," Zuckerman told me a few days after the streets finally cleared. "As I sit down and analyze the Twitter feeds, I've found one Moldovan in the U.S. who posted more than a thousand messages over the course of five days."

To Howard Rheingold, it matters little whether the Moldova protesters constituted a proper Twitter Mob. ("Enlisting the support of ex-pats and others around the world — keeping it alive via Twitter — was maybe more important than the coordinating function, the smart-mob function.") What matters is that the single unifying element (not counting entrenched corruption or tyranny) has undergirded every democracy struggle in the new millennium. That element is technology in the hands of the individual.

In the world of Listenomics, revolution has been revolutionized.

Chapter 11

NOBODY IS SAFE FROM EVERYBODY

CHAOS. SOMEDAY IT WILL END, and when it does, when things are sorted, when order has been restored, surely that will mean daisies and bluebirds and the laughter of children everywhere all the time, eh?

Nah. The very nature of Listenomics is such that there will always be some level of chaos in the scenario.

So, focusing as I have on all the magic and wonder and utopian promise of digital connectivity, perhaps I should end where I began: the darker side of progress. This, too, is about the infinite connections the digital world has established — connections that are difficult to sever and, in the ultimate free market of ideas, impossible to control. And it comes to this: in Society 2.0, you can run, but you cannot hide. When everybody is listening to everyone else, whatever you used to think of as your personal privacy cannot be guaranteed. Nor even expected. This is about how vulnerable you are. Not, "you," part of the society at large. You, personally.

Not at the hands of the U.S. Government or Procter & Gamble or some other looming Big Brother that may threaten you, by losing your personal data or selling it to insurance companies and other evil third parties or gaining access to your surfing habits to peer creepily into your life — though these are obviously legitimate concerns. As society reckons the trade-offs between absolute privacy and the incalculable benefits of a data-rich universe, protection from misfeasance, malfeasance and outright infofascism will require an evolution of law and public policy. Nobody wants an Orwellian future. On the other hand, while those issues are critical in principle and practice, the chance of harm they represent to a given individual is really quite remote. In a connected world, the problem isn't Big Brother. It's a billion little brothers.

"The Orwellian literary landscape," says Michael Fertik, founder of the software firm Reputation Defender, "imagines a universe where there's a concentrated amount of power in a central authority figure. Central authority is the hub around which the rest of us are spokes. Clearly, the model for the info universe has changed since Orwell's time. We now see that the world is one in which there are many hubs, many spokes." And many of those spokes aren't particular knowledgeable, or cautious, or honest, or fair, or kind, or civil or even humane when they sit at their keyboards. So, no, none of what you are about to read is fear about institutional snooping or hypothetical leaks. It's about your friends, your family, your neighbors, your colleagues, your classmates and a few billion total strangers who — thanks to the miracle of the internet — can make your life a living hell. You have very little to fear from *1984*, but every reason to quake about *Lord of the Flies*.

In this chapter, we'll look at a dozen cases, in escalating degrees of horror, of individual victimization. They vary in scale and in particulars, but they have one thing in common: the devastating combination of digital connectivity and a bottomless global reservoir of malice. Let us begin in this final chapter, as we did in the very first, with a bit of travelogue metaphor. Unlike our visit to coastal Montenegro, this one offers no mournful cello music. But it does offer the same echoes of Soviet Russia. It's a cautionary tale — set in, of all places — Estonia.

With Enemies Like This ...

As Larry King might say, when you look up "tiny Baltic nation" in the dictionary you'll find a picture of Estonia. It has a population of 1.3 million, sparsely settled in territory the size of Vermont and New Hampshire combined. Half the country is forest. It has a friendly rivalry with neighboring Latvia, but is feared by absolutely no one. The Estonian army has no tanks. The air force has two planes. If invaded by another country, the nation's defense strategy is literally to hide in the woods. Yet in 2007, Estonia found itself under attack — a successful attack that brought the country, if only briefly, to its knees. The invader, as far as anyone can tell, was Russia. The weapon was digits.

The hostilities came in several waves, a few exploratory raids followed by a major assault. The first attack was at 10:30 p.m. on April 27 against some government and media websites and an internet service provider. These were so-called Denial of Service attacks, in which thousands of computers from around the world were marshaled — most of them hijacked by hackers via downloaded worms — to flood the target servers with data, overwhelming their capacity and sending them crashing down. In some places, this might be regarded as a minor annoyance; servers crash all the time. But Estonia happens to among the handful of most wired — and internet-dependent — countries in the world.

"I park my car with my mobile phone," says Raul Rebane, a colorful cyber-warfare consultant (and, I kid you not, discus coach), who talked with me over coffee in Tallinn about national security and world-class steel-disc launching. "I haven't been in my bank for eight years, because I don't have to. We do our elections [online] with ID cards. When you have a sophisticated infrastructure, you are a target. There is no need to launch cyberwar in Kandahar."

All right, but why would anyone — much less Russia, which has bigger fish to fry — want to do harm to Estonia? Nobody has definitively answered that question, but there certainly are juicy circumstantial clues. One is quite basic. As a defeated former superpower, a petrodollar-flush Russia is eager to once again be a global force to be reckoned with. But apart from ICBMs, its military consists mainly of rusted tanks and miserable conscripts who anyway have their hands full in Chechnya. Cyberwar, however, requires no armor and the subject clearly fascinates the Russian military-academic complex. Rebane says there are 700 books on the subject in print in Russian. But how to take it beyond the theoretical? If a country were developing nukes, as we've recently learned in North Korea, it would do a test. If a country were preparing to the cripple the internet in one or more enemy nations, why wouldn't it test that capability, too?

But wait ... Estonia? "Enemy nation?"

Well, maybe. To visit Tallinn is to understand. In many ways, the charming Northern European city resembles Helsinki, just a salmon's throw across the Gulf of Finland. The language is similar, the architecture is identical, but the histories are very different. Estonia was occupied by Soviet Russia before World War II, taken and ruled by the Nazis during the war, and annexed by the Soviet Union for the entirety of the Cold War. Though 30% of Estonia's people are Russian speakers, many of them have no citizenship rights and represent a bitter underclass amid an Estonian-speaking majority itself embittered by decades of occupation.

The mutual enmity found a perfect symbol that spring when the City of Tallinn moved a World War II memorial, known as the Bronze Soldier, from the city center to a military cemetery on the outskirts. The statue commemorated the millions of Soviet soldiers who died during WWII, or, as it is known in Russia, The Great Patriotic War. To Estonia's Russian speakers, for whom the Bronze Soldier was a cherished shrine, the change of venue was deemed an intentional slight, resulting in violent street demonstrations and caustic rhetoric. But it took place as scheduled on April 26, 2007.

The first cyber assault came the next day, followed by several more on the 28th and 29th. The most damaging attack, however, did not occur until 11 p.m. Tallinn time on May 8th. It shut down major banks for 90 minutes—a brief but crippling blow to Estonia's largely cashless economy. As Raul Rebane puts it, "No transactions. No milk, no bread, no gasoline."

Once again, nobody can say for sure that Russia was behind the attack, although Estonian President Toomas Ilves has said there is no other suspect. It's worth noting that 11 p.m. May 8th in Tallinn is 12 midnight May 9th in Moscow. May 9th is Victory Day. The most sacred day on the Russian calendar, it celebrates the end of the Great Patriotic War. What a fitting occasion for orchestrated e-malice. And what a terrifying example of how a connected world can put you in the crosshairs of forces unknown, and perhaps far away.

Hacked

Now then, we were speaking of individual vulnerability. Let us each of us explore our inner Estonia, taking as our first example one of my favorite individuals, a man I'd personally trust with my life. I speak, of course, of me.

If you Google "Bob Garfield" (and I don't know why you wouldn't; I do it every day) you will get between 70,000 and 100,000 hits. Number 1 usually takes you to my blog. Number 5 is my NPR bio. And the rest are all over the lot. But I'd like to call your attention to one item that for the past seven years has fallen here and there among the Top 20. It is an *ad hominem* attack on yours truly, in which the most flattering descriptive is "boring, self-indulgent hack." I take exception to this representation, of course, because in my opinion I'm not that boring. But let's just call it an occupational hazard. The screed was posted by a disgruntled story subject — disgruntled in particular because he was one of a half-dozen interviewees for a 10-minute radio piece and not the centerpiece of a fulsome profile dedicated entirely to his worldview. Had he received, say, 9 ½ minutes of airtime, presumably he would have been a lot more gruntled. I won't trouble you by detailing his wild misstatements of fact, general cluelessness and paranoia. The salient point here is: Top 20 on Google!

Maybe it stays there because, ugh, lots of people are clicking on it.

Or maybe it's because my Bizarro Boswell is somehow gaming the algorithm and forcing his page higher in the search results. Either way, whenever someone from my worldwide cult of devotees, or a potential lecture client or my mother-in-law Googles me, there on the first page sits a most vivid character assassination titled "Bob Garfield's Boiled Soul." How lovely for me. By the way, the particular obsession of the guy who has invaded my lonely corner of cyberspace? Technology's assault on personal privacy. That strikes me as fairly ironic, but I shouldn't whine too much, because my problems are nothing.

Think for a moment instead about Bill Broydrick. He is a powerful Wisconsin lobbyist, who counts among his clients Aurora Health Care, Milwaukee Public Schools, Milwaukee Metropolitan Sewerage District and We Energies. But while he is well known in local legislative and political circles, he's pretty much anonymous to the public at large. Or, at least he was till August 2007. That's when his first Google page lit up with — at item Number 4 — "DC Madam's phone list linked to Wisconsin Political fugure [sic]."

The DC Madam, you may recall, was Deborah Jeane Palfrey, who was convicted of running a call-girl service in Washington that allegedly got inside the belt of various inside-the-beltway luminaries. Irritated at taking the fall while her clientele skated, she released her phone records going back years — records which, if diligently vetted, would reveal the identities of, ahem, escortees. One of them, former Eli Lilly CEO Randall Tobias, was forced to resign from his high position in the U.S. State Department. Another, Louisiana Senator David Vitter, had to publicly acknowledge his "sins."

Both of these guys were quintessentially legitimate targets. Vitter, a conservative Republican, was sanctimoniously flogging family values while paying for extramarital female companionship. Tobias, as President Bush's "AIDS Czar," had forcefully denounced prostitution — and, in politically-charged contravention of public-health consensus, advocated abstinence over condom use. In other words, they were both public officials whose official rhetoric was belied by their personal conduct. In still other words, they were hypocrites — on our dime. So, yeah: fair game.

But what about Bill Broydrick? He is not a public official, nor is he even exactly a public figure. Yes, lobbyists do influence public policy, but they also, more or less by definition, operate as private parties behind the scenes. By what measure does a lobbyist's private conduct rise to the level of public interest? Absent evidence that he was using Palfrey to pimp for the officials he was lobbying — and none has been put forward — what is the newsworthiness of a businessman with a sketchy sex life? I can answer that question: apart from pure titillation or *schadenfreude*, there is none.

Dialing for Diddlers

Cold comfort to Bill Broydrick, who was outed by an enterprising local TV reporter using an extraordinary internet tool. The reporter found it at DCPhonelist.com, a site developed by a couple of Harvard web geeks who realized that the data in Palfrey's phone records — essentially, thousands of phone numbers — could be mined if there were a specialized engine for searching them. They created just such a thing, enabling anybody to type in a phone number for cross check. If it appeared anywhere in Palfrey's records, it would register a hit, including times and dates. Broydrick, unfortunately, had a whole mess of them. The reporter, (from an NBC affiliate with the jaw-dropping call letters WTMJ) got a perfunctory no comment from Broydrick and rushed online with his scoop. Yep. A private citizen can't keep his pants zipped. Stop the presses.

The point here, though, isn't that one Milwaukee TV station has a low threshold of scandal. The point is that the scenario was being played out many times over, by wives, girlfriends, colleagues, rivals and maybe even extortionists who need only to click on a website to dig for dirt. Broydrick may have been cautious enough to elude detection by his wife and law partners, but nobody is safe from everybody. And in the digital world, everybody can get into your business — legitimate and unsavory — and cause all kinds of havoc thereafter.

"Exactly the sort of thing we'd hoped would happen," Daniel Silverman, co-creator of DCPhonelist.com told me after Broydrick's name turned up. It was vaguely creepy to hear that sort of gloating from him, because he is an earnest and thoughtful young man. But it's not so hard to understand. In Chapter 10, you already learned about Rathergate, and about Josh Marshall's Talking Points' deputization of citizen data miners to plow through thousands of Justice Department and White House emails in the fired-U.S.-Attorneys scandal. DCPhonelist, of course, employs the web in approximately the same way: crowd-sourcing a function historically dominated by a relative handful of journalists — journalists with a tiny fraction of the overall time and resources enjoyed by the whole blogosphere.

Still, what about relevance, and what about proportion? I asked Silverman if the highest and best use of journalistic crowd-sourcing was dialing for adulterers. He couldn't see why not.

"In this case," he said, "you may not think that what we're doing has as much journalistic integrity as Karl Rove's emails or of Enron documents, but I think it's just a first step. And I think it shows that this is going to keep happening, and every time data is released there are going to be more and better tools developed to analyze that data, for anyone to analyze that data. And so, while this data may not reach the level that some people consider to be highly important, I think that the philosophy behind it and what we're trying to do is very applicable to a wide range of issues." Data doesn't hurt people, he insists. People hurt people.

Not such an outlandish rationalization. Silverman is choosing to focus on the greater good, and not on what is in essence a parallel to the gun debate — i.e., don't blame the weapon for the weapon's misuse. Actually, an even better parallel is the Alfred Nobel problem. In 1867, Nobel invented dynamite, the first stable high explosive, which revolutionized civil engineering. Highways, dams, tunnels — the infrastructure of the industrial age — would have been primitive and limited without Nobel's discovery. But so would warfare and terrorism — the other beneficiaries of his ingenuity, a fact that haunted Nobel to the day he died. DCPhonelist.com is dynamite in the wrong hands — in this case because of the private nature of the behavior being investigated. The scale of the internet, combined with the anonymity and blind righteousness of the private dick-o-sphere, exposes all of us to not only to the curiosity of the crowd, but also to the malice. And there is no shortage of malice. Just click on the comments section of YouTube, or any sports or political site. The degree of bile and downright viciousness is simply phenomenal. Listen to this:

Filthy, dirty, lowlife, scumbag-ette, whore, slut, skank,...

Is that about Lucretia Borgia, Leni Riefenstahl, Medusa? No. It's about Sharon Stone, the fading film diva. And here's another lovely example of measured, respectful online discourse: *I wouldn't waste a fart in your general direction you fucking shit stain.... By the way pussy boy, what luckless teem are you backing? While the game is on today why don't you go outside and play hide and go fuck yourself.*

Saddam Hussein is in Good Company

Here again, the lyrical disquisition isn't about, for instance, a purveyor of neo-fascism. It isn't about child rape, or slavery or cannibalism. It's about the supremacy of the Dallas Cowboys. But what makes it so remarkable isn't that someone can get so angry and mean over something as trivial as a football team. There is no news in the pathetic fact that sporting passions run high, nor in the idea that morons have computers. There is, however, news in the fact that the internet essentially removes all barriers to hate speech. In the absence of social disincentives — identification, frowns of disapproval, litigation, punches in the nose — the anonymous online critic feels free to surrender to impulse, venting his rage, frustration and impotence in a place where his anguished voice can be heard. If the internet is indeed the "information superhighway," these guys are the ones in the '86 LeSabres, ruddy with Bondo, weaving in and out of traffic and cutting you off at 93 mph. Nor, in their seething discontent, do they seem to be especially discriminating. The "online disinhibition effect," it's called.

Adam Joinson, a professor of psychology at England's University of Bath, describes why comment boards and other online venues are such playgrounds for the id.

"Some research we did out of my lab found that when people are communicating online, they tend to become more focused on themselves, which means they're more focused on their attitudes and emotions," he says. "And if you combine that with a lack of concern about the person you're talking to, not being aware of their reaction and their response, then quite often you can get a powder keg where people do tend to vent, they do tend to flame."

For instance, just to amuse yourself, Google the phrase, "is such an asshole." You will get 34,000 hits. The targets include the obvious (George W. Bush, Barry Bonds, Paris Hilton, Donald Trump, "my boss," "my dad," Saddam, James Lipton), the less than obvious (God, Santa) and the truly unexpected (the sun, "my 3-year-old"). But, of course, many a just plain John and Jane Doe is on the list, too. The name may mean nothing to you scrolling this search. But if you happen to be Googling poor John or Jane, you will be getting some information that they'd probably prefer you not to have. And when they Google themselves, well, from personal experience I can attest that the exercise can be dispiriting.

"Rarely is the content itself what does the harm," says Michael Fertik, of Reputation Defender, which sells various monthly services to individuals and businesses seeking to monitor and/or burnish their web-utations. "It's often the content plus its positioning in Google. And studies have shown that the top number of results in Google, whether it's two or five, dominate a person's impression of the subject that they're searching. And they also don't, typically, dive very deep."

The particular research to which Fertik refers was done by the Sacramento, California eye-tracking/mapping firm Eyetools on readers of Google search pages. It demonstrated a "golden triangle" of 100% attention on the first three search results, dropping to 85% for the fourth, 60% for the fifth and only 20% by the 10th. This study was the basis of much academic hand-wringing over whether the Google algorithm — far from widening our information horizons exponentially — actually makes us cumulatively more ignorant by constantly recycling the most-linked-to results, in effect perpetuating intellectual shallowness. The answer is probably "yes and no," an ambiguity that does not apply on matters of online embarrassment. If the Google search pops up with "self-indulgent hack" or "brothel customer" or "Al Qaeda sleeper" in the top three or four results — whether they're true or not — you're just plain out of luck. As Fertik cautions, "If someone says about you that you've got herpes, the people who are going on a date with you or are considering a date with you are going to think twice because, even though the source may not be *The New York Times* — it may be something that is much less obviously reliable — they're still going to have smoke in the back of their minds that promises the threat of fire. So, first impressions are regrettably very important."

The phenomenon, of course, will disproportionately bedevil celebrities and anyone who does blunders into Warhol's 15 minutes of fame. Antonella Barba was an architecture student at Catholic University in 2006 when she made a big splash on *American Idol*. Described by the *Washington Post* as "a coltish beauty with a jazzy vibe," she was briefly the most Googled woman in America — though this perhaps owed less to her jazzy vibe than jazzy photos of her, along *Girls Gone Wild* lines, that were posted online. Shortly thereafter, she was eliminated from *Idol* and tried to return, at least temporarily, to obscurity — as she told the Post, "to stay out of the public eye a bit." In the winter of 2007, actor Alec Baldwin, a divorced dad who had been shabbily treated by his teenage daughter 3000 miles away, left a scorching voicemail berating the child for cruelty and disrespect toward her own father. The message found its way to the internet, exposing him to charges of child tele-abuse. Actor and screenwriter Eric Schaeffer wrote a book titled *I Can't Believe I'm Still Single*. This unleashed a torrent of online testimony from women far less incredulous than Schaeffer — including email traffic purportedly between Schaeffer and various of his previous dates. Now anyone who Googles him will find chapter and verse of narcissism and sexual perversion. His number 2 Google entry is headlined: "The World's Worst Person."

Obviously, Eric Schaeffer is not the world's worst person. He's at least dozens of places behind Kim Jong-il, Robert Mugabe, Bill O'Reilly and Osama bin Laden. And, one has to admit, Aleksey Vayner.

He was the Wall Street job candidate who sent a video resume in 2006 to the investment banking firm UBS. The video was so appallingly and hilariously self-aggrandizing, someone at the firm couldn't resist passing it along. The recipient also couldn't resist, and soon Vayner's vanity was a viral phenomenon on YouTube. "Impossible is nothing," he offered by way of his philosophy, borrowing from the Adidas shoe ad campaign of the same name, much as he apparently borrowed whole passages in his self-published book on the Holocaust, and as he borrowed the website template for his nonexistent hedge fund. The true beauty of the Vayner video, though, was catalogue of his world-class sporting achievements: his 140 mph tennis serve, his 495 lb bench press and, of course, his karate destruction of seven ceramic bricks. The downhill skiing shots (allegedly of somebody else) and ballroom dancing sequence with the totally hot babe with the bare midriff were also quite impressive. As one poster on one blog summed up the guy's prospects: "This guy is just a joke, and Google shall always remember his name as such." His name, face and achievements brick smashing will forever be associated with malignant self regard. He is, in short, an international laughingstock.

And it's not fair.

Yes, Vayner made a pompous ass of himself, and, yes, he's just the kind of self-inflated blowhard everyone loves to see get his comeuppance. But he also is a victim. His video was sent to UBS as privileged communication in what should have been a confidential personnel process. The leak was a gross invasion of his privacy, and one from which he will never fully recover. Fifteen years earlier, had a Yale student sent an identical resume to Wall Street, he also would have been a laughingstock — but the laughs would have been limited to a handful of people within about one square mile.

If Only it Had Been an Actual Light Saber

For the sake of argument, though, let's just say Vayner and Schaeffer and Baldwin and Barba were victimized substantially as a result of their own behavior, that somehow they had it coming to them. The fact remains that these are the sorts of trespasses that hitherto have not become gossip fodder for the whole world. In a world of camera phones and MP3 files and TMZ.com amd MySpace, it's best never to do anything ridiculous. Ever. Whether you're a celebrity or not. Whether you're a jerk or not. Whether you're an adult or not. On that subject, perhaps the world's foremost expert is Ghyslain Raza, the Quebec teenager you know as the Star Wars Kid.

In 2003, Ghyslain went into his school's media room and taped himself mimicking the character Darth Maul's "light saber" sequence from *Star Wars Episode 1: The Phantom Menace*. He says he was choreographing a bit for a student video he was producing, but that hardly matters. What matters is that he left the tape in the camera, and it was later discovered by a classmate — who was struck by how unlike Darth Maul the chubby Ghyslain appeared, especially since the weapon was not actually a light saber but in fact a telescoping golf-ball retriever. That kid showed it to another kid, who showed it to another, who converted it to a .wav file and posted it online. Six years later, Star Wars Kid has been seen online — and laughed at — more than 1 billion times, making Ghyslain undoubtedly the most ridiculed among all the ridiculed fat kids in human history. According to depositions in a lawsuit filed by his family against his classmates, the boy spiraled into a depression and was forced to leave school — where, when he entered the cafeteria, fellow students had been shouting, in unison, "Star Wars!"

He has also been the constant victim of cyber-bullying, mainly focused on his weight. Even after the boy's suffering had been widely reported, he generated astonishingly little sympathy online. One anonymous, and sadly typical, poster had this to say in 2006 after the lawsuit was settled: "Ironically, for me, knowing the kid is actually a whiny little bitch makes it even funnier to see him dance around like an idiot. Instead of kinda identifying with him in doing something a little goofy, I now just think he's a douche bag and deserves to be the laughing stock of the internet."

Don't be especially shocked by the epithet. To this day, if you Google "Star Wars Kid" and "douchebag," you get more than 13,000 results. (For more of your douchebag-identification needs, consult douchebag. com, which lists a handy "Douchebag of the Day.") It would be nice to shrug that off, and say "He's young. This, too, shall pass. Get back up on the horse, kid. Time heals all wounds." Would that it were so, but such commonplaces do not necessarily apply to the digital world. It has often been observed that Google is God, which may or may not be so. We do know this, though: sometimes it is benevolent, and sometime it is wrathful. Verily, Google giveth, and Google taketh away. Thanks to the search engine's endless cache, internet wounds last forever.

"My clients wait for something to go away. Then they find out it just doesn't,." says Nino Kader, founder of International Reputation Management, a Washington, D.C. public relations firm dedicated, like Fertik's Reputation Defender, specifically to the realities of the internet — "which in my opinion," Kader says, "is going to become the legacy of record for anyone who ever existed on the planet."

Exactly. Never mind what Andy Warhol said. In the future, everybody will be slandered in perpetuity.

Remember when you were in school? Remember what Miss Dingler told you, to discourage you from cheating or throwing iceballs at the 4th graders? She said, "It will be on your Permanent Record," and that was something to think about. Some day you would be trying to get into the military, or trying to avoid the military, or applying for a job, and someone would be looking over file with a scowl on his face. "Well, it says here that you were caught passing a note to Philip Yampolsky about Jane Konowich. Is that correct?" And that would be that. Your troubled history would have finally caught up with you. Your West Point appointment and pitching career with the New York Yankees would be over before it ever began. It was a scenario that haunted me — along with nuclear holocaust and swimming with a full stomach — from the age of 6. So intimidated was I by the specter of my Permanent Record that I toed the line in school for 12 years. OK, maybe there was the slightest bit of *World Book Encyclopedia* plagiarism and chronic absenteeism and perhaps a touch of substance abuse, but I stayed under the radar — or, at least, so I assumed. A few years ago, with some trepidation, I telephoned the Lower Merion School District to discover, at long last, what was indelibly inscribed in my dossier.

> *"Your permanent record?" said the voice on the phone. "That's been shredded."*
> *"Shredded?!" I yelped. "Shredded when?"*
> *"When did you graduate?"*
> *"1973."*
> *"Well, then, 1974."*

So much for permanence. If I'd known they were bluffing, there might have been a lot more mischief in my past. Alas, the ability to shred the unpleasantness is in the past, too. Those internet pages are cached till doomsday — the world's or yours. The descriptions of my boiled soul and Bill Broydrick's secrets for meeting hot women are not going anywhere. On the contrary, they are proliferating, because of the ever-expanding blogosphere. "Before blogs," Kader says, "you had to start a website — with all the terms and conditions, so there was some accountability. Not much, but some." With do-it-yourself blog tools, "you could publish to the web anonymously within a matter of minutes." And if you wanted to call Santa Claus an asshole, well, tough shit for Santa. "Anybody can anonymously bash anybody else — and it will be available for the world to see."

The knowledge of which is always nerve-wracking, sometimes nightmarish and occasionally tragic beyond words.

Remember Megan Meier.

She was a 13-year-old girl from a suburb of St. Louis who made a fast friend on MySpace, a sweet and cute 16-year-old named Josh Evans. Over a period of weeks, they chatted and flirted with one another, sharing favorites and adolescent intimacies and the contents of their teen souls. Then, out of nowhere, Josh turned on her. He called her horrible names. He ridiculed her. He told her "everybody hates you.... The world would be a better place without you." Then others in the social network joined in the free-for-all of bile and recrimination, which ended only when Megan, just shy of her 14th birthday, committed suicide. It was a story that resonated, because though the cruelty of children is as old as the ages, the e-bullying angle made it a sign of the times. Or so it seemed until the narrative took a shocking, unimaginable turn. There *was* no Josh Evans. He was an invented character, created not by another child, but by two adults — one of them the mother of one of Megan Meier's ex-friends. The mother, Lori Drew, explained that Megan had spread gossip about her daughter, and she wanted to "mess with her." She succeeded.

You've Got to Ac-cen-tuate the Positive ...

So, in a digital universe that leaves anyone and everyone so exposed, what is there to do? There is no single answer, although one good one might be to never do anything — anything — that you don't want to see pictures of on your tween's cell phone. Another concerns shutting the barn door after the cow has fled, or, as the practice is also known: public relations. Image management has long been the mainstay of politicians, celebrities, business executives and other notables dissatisfied with how they are portrayed in the media or perceived by the public. Recently, a document surfaced from one such prominent American, who in 1970 obsessed over the fact that his image as a humorless stiff and all-around SOB did not reflect his softer side. So he knocked out an 11-page memo on his inner sweetness, asking his staff to get the media to recognize, for instance, his "treatment of household staff, the elevator operators, the calls that I make to people when they're sick, even though they no longer mean anything to anybody, the innumerable letters I have written to people when they have fallen on bad days, including even losing an election. *No president could have done more than I have done in this respect, and particularly in the sense that I treated them like dignified human beings and not like dirt under my feet.*"

The italics are mine, because the phrase is just so … italicizable. Poor Richard M. Nixon. So tragically misunderstood.

The impulse to promulgate revisionist history, or to put your best foot forward, or simply to correct the record was not unique to Tricky Dick. Nino Kader's clients at International Reputation Management have exactly the same idea. But, unlike the late president, before they can address their media profile they have to deal with their Google one.

"They want to highlight all the good things about them," Kader says, "so they're pushing the negative right off the page."

That may represent a Sisyphean task for someone like Aleksey Vayner, who still generates 20,000-some search results, all of them bad. But most of us are, shall I say, less image-disadvantaged. One of Kader's clients, for example, is a prominent TV talking head on a very polarizing subject. He was dismayed to see his Google page pop up not only with nasty speculation about his political motivations and loyalties, but also about his physical appearance — to wit: "If this is the same [pundit's name] that I've heard pop off at the mouth on MSNBC and FOX News, perhaps a weight reduction program would be appropriate to reduce the fat surrounding his brain in an attempt to obtain a more clear, rational thought process. I've seen his chins gyrate far too many times."

Ouch. Bad enough to be called irrational, but "fatso" really smarts. It's harder to be taken seriously as a commentator on geopolitics when you are perceived as the Star Wars Kid of public policy. So Kader got busy. "I created an official bio of him," mentioning nothing about his chins, but rather described his "Sensible, Skeptical" analysis of the day's events. "It appeared number one on Google." And, for a while, it worked, pushing the double chins off the main page. Unfortunately, it's a jungle out there, and the jungle is persistent. Without constant effort, all attempts at cultivation will be overwhelmed by the sheer irrepressibility of nature. Kader's client eventually discontinued his website, and the corpulence issue is back on the first page.

The concept, however, is proven. Creating pages and links to those pages, and making sure they are maintained, can, as Kader says, "bulletproof your image on the Web. For example: Hillary Clinton. She's got the first two pages covered."

Yes, although she is among the most despised political figures in America (and, yes, I know, also among the most popular), to Google candidate Hillary in the summer of 2007 was to find her entry on Wikipedia, her official campaign site, her U.S. Senate site, her MySpace page, the White House page on her tenure as First Lady and a *New York Times* profile. "If somebody wanted to say she stinks," Kader observes, admiringly, "it would be difficult to punch through." For instance, scrolling through the search results, you can go through at least a dozen pages and not see the word "Whitewater."

Some of that is luck, and some of it is technique. The problem is, knowledge of the technique isn't limited to your friends. The author of "Bob Garfield's Boiled Soul" knew exactly what he was doing. "He called the page 'Bob Garfield,'" Kader explains, "so Google sees that as relevant.. The headline is in large font, too. If you bold it, Google thinks it's important. That's why he did it. He also has 225 links to his site. Google will consider a link into a site as a vote." The same gimmick was famously employed to link the Googled search terms "useless failure" directly to the official White House profile of George W. Bush.

Kader takes pains to point out that he does not provide his service to just anybody. "There's a doctor who contacted us. He botches a lot of surgeries, or something. I don't think that was something I wanted to suppress." Alas, the rest of the world is not always so discriminating. Among the things that the free market of ideas is free of is conscience. That's where, every so often — and none too efficiently — the law comes in.

The Megan Meier case was one such. Because Missouri has no criminal statute applicable to cyber-bullying, local prosecutors were helpless to press charges. So a California U.S. Attorney took it upon himself to prosecute Lori Drew. For computer hacking. Pinning his legal argument on the fact that Drew technically violated MySpace's terms-of-use agreement by creating a false identity — obviously common practice throughout the internet and social networks in particular — the federal prosecutor obtained a conviction on three misdemeanor counts of illegally entering MySpace. A dangerous precedent if ever there was one.

Christopher Soghoian, of Harvard University's Berkman Center for Internet and Society, points out that this use of the hacking law exposes millions upon millions of internet users to prosecution. "Google's terms of service," Soghoian says, "specifically forbid anyone under the age of 18 from engaging in a Web search, from sending an email, from uploading a photo to their Picasa website. MySpace's terms of service specifically state that you may not, as an end use, upload a photograph of another person that you have posted, without that person's consent. What this means is that if you take a photo at your family reunion, you must go around the room and obtain every single person's permission to upload that photo to your MySpace page later that day. If you don't do it, you're a criminal."

The decision was immediately controversial and was widely expected to be dead on arrival in federal appeals court. Among the critics was Stanford University law professor Mark Lemley, who believes that — absent evidence Drew was trying to drive Megan Meier to suicide — this was not a matter for the criminal courts. "I think there's a perfectly good civil case here for wrongful death, for intentional infliction or for reckless infliction of emotional distress," he says. "The parents of the teenager may well have a cause of action against Lori Drew. But that's a different matter than saying she needs to be in jail for the crime of pretending to be someone she wasn't."

Uncivil Liberties

On that subject, Lemley has more than passing familiarity. He represents two Yale Law students — Brittan Heller and Heide Iravani — who were repeatedly, viciously targeted by anonymous online critics on a law-school admissions discussion board called AutoAdmit.com. I see no need to repeat the verbatim calumnies here. Suffice to say the women were accused, in the most vulgar terms, of everything from sexual impropriety to heroin addiction to cheating, and were subject to lascivious speculation of the most graphic and sometimes violent kind by dozens of unknown character assassins. What began as a post ostensibly warning Yale law students that a "bitch" named Brittan Heller was soon to be among them quickly triggered a fusillade of increasingly hateful potshots from persons unknown. Iravani's nightmare began later, when she was targeted with repeated obscenities, and much worse, for the crime of being shapely.

"We've got a pretty broad protection in the United States for speech that expresses opinions, even offensive opinions about matters of public concern and about people," Lemley says. "But, you've got no right to tell lies about people."

Nor was this merely a matter of humiliation. As David Margolick wrote in the March 2009 *Portfolio*, Heller says she was turned down by the first 16 law firms she's approached for job interviews, because — she was certain — a perfunctory human-resources Google search returned results from AutoAdmit besmirching her every which way. Separately at first and eventually together, Heller and Iravani tried various means of undoing, or at least minimizing, the damage. They asked AutoAdmit to remove the offensive postings, but the proprietors refused, citing freedom of speech. They hired Reputation Defender, which started a petition drive to shame AutoAdmit into policing its site. The result was still more online hostility from commenters incensed over the supposed coercion and threat of censorship. The women asked Google to take down the defamatory posts from its search results, but Google also refused, citing its own policy and the explicit protection of publishers, search engines and internet service providers under Section 230 of the Communications Decency Act.

"That immunity provision doesn't have a counterpart in the physical world," Lemley says. "It was created because of one of the significant differences between the internet and the physical world, which is the sheer volume of information that can be posted on the electronic equivalence of bulletin boards"

Not that he necessarily objects to the provision. He knows what Congress had in mind: 1) the technical impossibility of monitoring every online utterance for truth, much less civility, and 2) the understanding the any legal liability to various online hosts would result in draconian measures toward self-protection, with a consequent chilling effect on free expression. Thus Congress offered blanket immunity to everyone but the author of libelous or defamatory content. If you've been slimed online, and you want justice, you've got to take it up with the slimer. "So it's not that you're without recourse," Lemley says. "It is, however, the case that you've got to find and track down the people that are actually doing the posting online, and that can sometimes be difficult."

Nonetheless, Heller and Iravani attempted just that, and managed to track down five of their 40 most abusive tormenters and sue them for defamation. They haven't located the worst culprits, but the women have gotten the satisfaction of putting a handful of no-longer-anonymous law students in the same glare of opprobrium and job-search jeopardy the women faced themselves. They might have been able to learn the identities of the others, but AutoAdmit — under the protection of Section 230 — had scrubbed its servers of the IP addresses that would have helped the plaintiffs locate the authors.

Shouting Whore in a Crowded Internet

So should the law be changed, to protect the Brittan Hellers and Megan Meiers and even Bob Garfields of the world? (Just in the past week, as I write this, I've been accused online both of taking bribes from Comcast and inciting people to *murder* Comcast employees.) The Communications Decency Act, after all, was passed in 1996, while the internet was in its infancy.

"That statute is not the model of clarity," says Kurt Opsahl, senior staff attorney for the Electronic Frontier Foundation, "Perhaps it is a little outdated, but I wouldn't support a change to it that would provide less privacy and less protection for free speech." Opsahl offers several specific reasons, but not before quoting the early 20th-century philosopher-jurist Learned Hand's meditations on free speech: "'Right conclusions are more likely to be gathered out of a multitude of tongues, than through any kind of authoritative selection. To many this is, and will always be, folly; but we have staked upon it our all.' I think what he meant by that is that the marketplace of ideas requires a vibrant marketplace and some of those voices may be useful, many of those voices you might disagree with. But we as a society have taken a bet that it would be better to have a lot of different voices rather than have the law or the courts decide which voices can come forward."

This country has always had rude, *ad hominem* speech, Opsahl observes, clear back to the founders. Moreover, the public has always been able to sort out the sane voices from the deranged ones, whether on a 17th century pamphlet or a highway overpass or a men's room wall. "A lot of these sites, if you look at the context in which they are written, it's hard to find them very credible. You know, if you see somebody say that so-and-so is a criminal on a site that is filled with hyperbole and bile, how much credence should you really give it? Not much." Furthermore, he says, there is the self-correcting aspect of the internet, because false and scurrilous comments can quickly be answered to balance the record.

Yeah? Tell that to Google, which reports the results most linked to, not the ones containing the most truth, and certainly not in their original obviously unsavory environs. Ask Brittan Heller, who sued her tormentors, hired Reputation Defender and still must live with a Google results page that has, at or near the top, "Is Brittan Heller a Lying Bitch?"

To Reputation Defender's Michael Fertik, this screams for a revision to the law, giving the targets of character assassination at least as much protection as the Digital Millennium Copyright Act of 1998 confers upon intellectual-property owners. "Under that law," he says, "a website is not liable for the content that anybody puts there, including copyrighted content, until it receives notice that that content is copyrighted. And then it becomes liable. That's called a notice-and-takedown provision — once they're put on notice then they have to take it down. Now, under this comparative set of regimes, NBC can send one letter to YouTube and force it to remove 50,000 videos, as happened." But if someone Photoshops you into a crime scene and makes you look like a brutal killer, then posts it all over the internet with your name and address attached … well, that's just hard cheese. Yes, under current law, a *Saturday Night Live* skit is afforded more protection than you are. Opsahl, of the Electronic Frontier Foundation, worries that giving privacy the same status as copyright would encourage anybody whose feelings get bruised — especially government officials — to inundate websites with demands to take down unflattering material, even truthful unflattering material. But Fertik says that's silly. He's not talking about wholesale revisionism.

"There's an onus on the person who's asserting the copyright ownership to prove it, to make a certain showing of proof. It's not very burdensome, but there is a burden that they have to carry. But it's not a two year burden through a judicial system, a John Doe lawsuit, a complaint that has to be filed and adjudicated over years — meanwhile your entire life has been taken apart by a website that's defaming you — it's a very different level of onus. So that's a pretty good system. We could require, for example, that anyone who believes himself or herself to be defamed to file an affidavit saying that to the best of knowledge this is totally false and to be penalized under penalty of perjury if his or her own claim, filed in front of a clerk of the court, turns out to be false."

Fertik also has little patience for First Amendment absolutists who see increased protection against defamation as creeping censorship and suppression of democracy. "It's not like the alternative to the status quo is totalitarianism. It's not true, it's false, it's wrong. There are 100 steps between where we are today and a scenario in which we really chill speech. The laws of the internet do make sense and did make sense when the internet was just coming up to speed. The internet has now grown up, we've become more mature, the internet has become more mature and so the law has to catch up, so I suggest we take three steps down that primrose path, not 100."

The Celebration of the Jackass

In the end, though, the statute book is thin protection against human nature, including, but not limited to anger, resentment, pettiness, feelings of impotence, meanness, cowardice and the absolute impunity conferred by anonymity. The internet is a playground for the id, which, as Sigmund Freud himself described it, is "a cauldron full of seething excitations. It is filled with energy reaching it from the instincts, but it has no organization, produces no collective will, but only a striving to bring about the satisfaction of the instinctual needs subject to the observance of the pleasure principle." Funnily enough, Freud had a handy synonym for the id: chaos. Anyway, it is alive and well on a website near you. Let's look for a moment at Dumponyou.com, ("Who do you hate and want to bash today?") which encourages users to "share your anonymous stories with the world."

You can "anonymously" post a story about a neighbor, co-worker, politician, or anyone else. Then you can send an "anonymous" email to other neighbors, etc., and tell them the story is there so they can come and read it. They can post "anonymously" any comments they have about your story. If it's a problem you'd like to discuss with the group, you can "anonymously" invite them into an "anonymous" chat room, where nobody can be identified.

Anonymity, needless to say, does not stimulate an excess of caution, or responsibility or, you know, evidence. Here's an example from the website:

I've sent anonymous emails to everyone in our sales group to give you a warning about Todd, our new VP of Sales. You all know he's Mr. Breslins nephew and that's probably the only reason he was hired. I'm sure you all could tell as soon as he was introduced to us that he is gay. Personally, I don't have a problem with that as long as he keeps his preferences to himself. Todd was in some real trouble for sexual discrimination and harassment in his last job. If you Google him you'll find out what I mean. Just put his first and last name in and Seattle, Washington, and all the headlines come up. He fired several people for no apparent reason and replaced them with gay friends. The whole mess ended up in suit with the company settling out of court and having to pay each of the people that got fired. Maybe Todd has changed his ways and will be different working for his uncle, but I felt I had to give all of you fair warning so you'd be a little more careful. I've also sent emails to Mr. Breslin and to Todd so they could read this and he'd know we are aware of his problem and we are watching.

The Dumponyou.com logo, fittingly enough, is a jackass sitting on a toilet.

"It's something I kind of thought would be a little bit different. It's more venting than anything else," says Howard Baer, the Scottsdale, Arizona, entrepreneur behind both Dumponyou.com and the equally mean-spirited TheAnonymousEmail.com (suggested message: "You're so ugly you need to sneak up on a glass of water to take a drink").

Baer — who until recently was a purveyor of mail-order immune-system supplements, and, according to the U.S. Securities and Exchange Commission, a "recidivist securities law violator" in manipulating his pill-pushing firm's stock price — contends that his services give a voice to the hitherto muzzled, those with legitimate gripes prevented from expressing them for fear of retaliation. But what about illegitimate gripes? What about spurious claims? What about intentional smears? What about innocent victims?

"That is a problem," he tells me. "These things happen. I think it's going to bring out a lot of bad, but it's not going to stop us from being in business. It's a business, and a business is a business whether it's good bad or indifferent."

Cloaking himself in freedom of speech, Baer is fond of issuing press releases casting himself as a first-amendment crusader. But, of course, the first amendment is a refuge for all manner of jackasses sitting on toilets. To get a bit of insight into the worldview — and ideological consistency — of this freedom fighter, here's an excerpt from his 2003 interview with Greg Tingle of the Australian website mediaman.com. Baer:

> *I think any American burning the American flag should be jailed. I think other countries that we support with food, money and military protection that burn the American flag or allow their citizens to do so should be shut off automatically. These people go out and burn the flag, burn mock ups of our presidents, then go home and eat the food we put on their tables. Bullshit, that is wrong. All Americans work their asses off for their wages and should not have their tax dollars support the animals in some of these third world countries.*

The flag-burning quote and SEC rap sheet, by the way, come courtesy of Google.

Unfortunately, Baer is hardly alone in his endeavors. Hate, revenge and anonymous flaming constitute an online industry, including, but by no means limited to, ThePayback.com, "for all your revenge needs" and the especially sadistic RevengeWorld.com, where spurned or otherwise spiteful men post pornographic photos of their ex-girlfriends. A related — and arguably more benign — category is the "women beware" genre such as Womansavers.com, a sort of ePinions.com devoted to alleged cads, creeps, brutes and philanderers. Women rate the guys they've dated on honesty, fidelity and conduct, presumably to alert their clueless sisters about the potential nightmares lurking in the dating pool. Needless to say, there are very few Mr. Rights catalogued here. And needless to say, readers are privy only to the woman's side of the story. But, of course, revenge can also be the private, noncommercial pursuit of the lone aggrieved, like my personal soul-boiler or Chicago blogger Craig Gunderson.

In June 2007, Gunderson ("I'm just so angry all the time," according to his blog profile) was peeved at having been flamed by an ex-colleague.

So he spent the day following links, IP addresses and other digital clues to gather as much embarrassing information on the flamer as he could, then posted it on his blog. This he deemed a lesson to the poor shnook he called "Huey." Gunderson:

> *So my point is the internets are the biggest, brittlest glass houses out there. It's wide open and that openess is there to keep us honest and in-check. There are no police in the interwebs, we are the police. Huey threw a rock yesterday and he should really know better.*

Knowing better. He raises an interesting point. After all, a good many of the "victims" of online abuse are more than worthy of the abuse they are taking. Herein lay the moral quandary of vigilantism. In a society ruled by law, courts are supposed to adjudicate guilt and liability. Mobs, by their very nature, are unruly and prone to error. They have little regard for subtlety and certainly offer no forum for the defense. But sometimes they catch the bad guy.

A Perversion of Justice?

The quintessential example is the aptly named Perverted Justice, the private organization whose members pose in online chat rooms as young teens, trolling for pedophiles. They then lure the adult target into a *tête à tête* with the "child." When the target arrives for the rendezvous, however, he is confronted by police and, sometimes, the rolling cameras of *Dateline NBC*. Should private individuals — not to mention news organizations — be in the business of entrapping fellow citizens who otherwise might never have acted on their impulses? Maybe not, but it's hard to argue too strenuously against child predators being exposed before they can do harm.

Nor, would it seem, could anybody persuasively argue against criminal defendants, especially poor ones without access to fancy lawyers and private eyes, knowing the motives — especially dubious ones — of those testifying against them. Our justice system has been forever blemished by testimony from unsavory characters given in exchange for judicial leniency, early parole or who knows what. The annals of criminal law sadly overflow with verdicts of guilt — even sentences of death — against the wrongly accused, based on perjured testimony tendered in a plea bargain or similar deal. Hence the rationale, or at least the ostensible rationale, for Whosarat.com, the "largest online database of informants and agents."

> *Who's a Rat is a database driven website designed to assist attorneys and criminal defendants with few resources. The purpose of this website is for individuals and attorneys to post, share and request any and all information that has been made public at some point to at least 1 person of the public prior to posting it on this site pertaining to local, state and federal Informants and Law Enforcement Officers. This includes an Informant who makes his or her Informant status known to any person.*

That's what it says on the site's home page, and it sound innocent enough, even beneficent. So why are prosecutors and cops — and presumably a good number of informants and agents — freaking out? Because on the same home page, right at the top, are displayed three mug shots and names of "Rats of the Week." Registering with Whosarat. com enables a defense lawyer, or a bailed-out criminal, or an angry relative, or a hit man, to access a complete profile of the "rat." While, as Daniel Silverman has argued, data may in fact be neutral, please note three salient facts:

1) While conflict of interest between the informant and the state clearly sometimes results in a miscarriage of justice, nobody has ever demonstrated that these cases are more than grotesque anomalies, and many a monster has been locked up based on the truthful testimony of an informant, rewarded with leniency or no.

2) Juries are already encouraged to consider the motives of all witnesses.

3) "Rat" is not a neutral word. On the contrary, the tone of this site

is so caustic and inflammatory that the proprietors felt obliged to post a disclaimer:

> *THIS WEBSITE DOES NOT PROMOTE OR CONDONE VIO-LENCE OR ILLEGAL ACTIVITY AGAINST INFORMANTS OR LAW ENFORCEMENT OFFICERS. IF YOU POST ANY-THING ANYWHERE ON THIS SITE RELATING TO VIO-LENCE OR ILLEGAL ACTIVITY AGAINST INFORMANTS OR OFFICERS YOUR POST WILL BE REMOVED AND YOU WILL BE BANNED FROM THIS WEBSITE.*

Duly noted. Threatening postings will be dealt with administratively.

And assassinations?

Pushing Buttons

Anger, jealousy, revenge, *schadenfreude*, ridicule, sexual harassment, condescension, malignant self regard. Online personal attacks cover the whole, vast, ugly spectrum of malice. But perhaps none so ugly as the modus of the online troll — namely, cruelty for its own sake. A troll haunts chat rooms, comment sections, discussion boards in the same way a pedophile hangs around schoolyards or a mugger waits outside of bars for drunks. They're on the prowl for vulnerable victims. Oh, there's no physical assault, but it's an assault nonetheless as, literally and figuratively, the troll contrives to push the right buttons to wound the victim. The act can be as banal as a political screed, intended to raise partisan ire, or an adolescent sexual slur, or a racial epithet or perhaps an outrageous lie. On the other end of the spectrum, it can be an organized campaign to torment a (usually) innocent party via all manner of vandalism, pranks, harassment, hate speech and worse.

"Trolling is messing with people who are easy to mess with, not because you have a personal gripe with, but because it's fun," says Mattathias Schwartz, author of a 2008 *New York Times Magazine* piece on the subject. "They do it because they don't empathize with their targets. They do it for a sense of release … to laugh as you watch them melt down." Trolling can be a solitary enterprise—some random jerk accusing you of inciting murder against the cable guy—but it is also an active internet subculture, largely congregated around the /b/ discussion board on the website 4chan.org, and the wiki site Encyclopedia Dramatica. There trolls gather to exchange ideas for what they call "lulz," a mutation of the online abbreviation LOL for laugh-out-loud. Their hobby: on- and offline mind games.

Schwartz, who spent time with ubertroll "moot," hoax artist Jason Fortuny and other malicious pranksters in reporting his story, says that the behavior is clearly sociopathic, but in some cases is at least rationalized by an underlying troll-deology: "that by inflicting emotional pain through the internet, 'we will teach them to be less sensitive. We'll teach them to put less personal information out there.' But this is just a justification, eloquently espoused by Jason Fortuny. They are manipulating people for their pleasure, and causing other people the maximum amount of pain."

None more than the grieving parents of Mitchell Henderson, who in April 2006 lost their seventh grader to suicide. In the days after Mitchell's death, person or persons unknown logged onto his MySpace page and learned that the boy had recently lost his iPod. This was promptly twisted into "Mitchell Henderson killed himself because of his lost iPod"—a notion that somehow captured the imagination of the habitués of 4chan's /b/ board. So naturally they started flaming the dead child, activity that got only worse when some well-intentioned defender of Mitchell's memory left a comment describing the boy as "an hero." This was simply too delicious for the lulzers, who conspired to make a teen's suicide into the funniest online meme ever.

"The further and further away it got from the actual suicide," Schwartz says, "the more people began to find it incredibly funny"—to the point that many eventually joined in the fun without knowing that Mitchell Henderson had been a real, live boy and not some sort of urban myth. But others in the cult were only too well aware that Mitchell death was a matter of actual flesh and blood. They sent pizzas to his parents, from their dead son. They phoned his parents, identifying themselves as Mitchell's ghost. "Did you find my iPod?" not-Mitchell asked. There were lulz all around.

To this, again and again, now for more than three years, the devastated family has listened.

AFTERWORD:
THE UNTOLD TRUE STORY OF *THE CHAOS SCENARIO*

IT'S JANUARY 2005. There is a meeting going on in a hotel conference room. About 50 people are in attendance, many of them hung over. One of them, bleary of eye and scant of attention, is me.

It's an *Advertising Age* editorial retreat. In my capacity as an ad critic — a columnist who keeps his job merely by filing 600 sparkling and perspicacious words once a week — my main role at the meeting is to hang out with my colleagues, crack jokes and carouse like an idiot. I have performed my job well. But here we are on the final morning, at the question-and-answer session where reporters are encouraged to raise issues concerning, make suggestions for and whine to editors about the day-to-day operation of the publication. As usual, I'm counting ceiling tiles. But, suddenly, a mental bulldozer pushes aside millions of dead brain cells to reveal an idea, which I share with the bosses.

"Hey, since we've done so much reporting on the internet, and problems in the media [in those days there was no Traditional Media. It was just media], and since it's pretty obvious everything we've ever counted on is crashing and burning, shouldn't we at least have some sort of logo on each story on those topics? Something like, "Chronicles of a Revolution?"

I'm not the only one running on fumes this morning. Evidently management is, too, because they hear my suggestion for a logo — you know, a little visual signature — and somehow get the idea that I've volunteered to *report* on the ongoing carnage. Then they assign me a story on the subject. My sweet lord, I will never drink again.

Anyway, about three months later, in April 2005, my first installment was printed in *Advertising Age*. It was titled "The Chaos Scenario." The industry went totally batshit, whereupon I resolved to do two things: 1) keep writing on the subject, about twice a year and in an AdAge.com blog dubbed "The Bobosphere." 2) package all of my reporting into an authoritative book, bringing me wealth and fame and the respect of my children. Specifically, the notion was to write the book online in full public view, so that I might have the benefit of my expert audience's experience, thoughts, misgivings and so on. What a swell plan.

Nobody paid the slightest bit of attention. Until two years later, when I took on Comcast, the traffic on my blog was near zero. But I kept at it, writing blog posts and stories, mainly for *Ad Age* but also one big one, on YouTube, for *Wired*. If you by any chance saw these pieces and posts, this book might trigger an eerie sense of déjà vu. Albeit not as much as I'd have liked. The other problem with my plan is that writing a book about the digital world is like trying to sketch the Kentucky Derby. Every time you look up, the picture has changed. For instance, when my story first ran — never mind Twitter or MyBarackObama. com — YouTube did not exist.

This is what I call "currency fluctuation," which is how a two-year project can expand to more than four years, as events constantly overtook my attempts to be current. It also had a fairly big effect on my publishing plans. At its inception, the book was meant to be published and distributed online until such time as a canny publisher swooped in to buy print rights and turn an online phenomenon into a print edition. My abject failure to quickly aggregate a worldwide cult of followers, however, led me to a more traditional solution: sell the book for an obscene advance to a traditional publisher, which publisher would be obliged to embrace the very marketing and distribution innovations espoused in the book. Not a bad Plan B, really, because a $150,000 advance, socked away, would guarantee my 8-year-old's college education with some left over for that Odyssey putter I've always wanted. So I engaged an agent and sat in horror as publisher after publisher after publisher responding with extremely flattering letters that ended by telling me to pound sand.

On the grounds that 1) the subject matter was moving forward too fast to document in a book, and 2) all this disruption stuff sounded so familiar. "Haven't we read this before?"

Well, no, not exactly. But never mind that. Am I not flesh and blood? Have I no feelings? Have I no pride? After many months of these responses, I sought the advice and counsel from friends, colleagues and near total strangers who have written on subjects digital along the whole continuum from traditional Famous Name Imprint to vanity press to no press at all. One of them was Greg Stielstra, a Nashville social-media expert, author of the books *PyroMarketing* and *Faith-Based Marketing*. I had read *PyroMarketing*, and seen him speak on his participation in the marketing and distribution of the mega-ultra best-seller *The Purpose Driven Life*. He was already aware of my project, and had actually been in touch to see how he and his employer, The Buntin Group, might promote my book as an illustration of digital-age solutions, while giving it the benefit of those very solutions. By the time I spoke to him in the first week of January, 2009, we'd long since agreed generally to collaborate. But by the end of the end of our January conversation, we were full partners in the present enterprise: employing the principles *of* Listenomics to publish this book *about* Listenomics.

In our attempt not only to publish this book to create a new template for publishing altogether, Greg recruited an ad hoc Team Chaos of web designers, social-media gurus, academics, online-buzz monitors, widgeteers, public relations specialists, video producers and video hosts to re-imagine book publishing. This effort, we believe, is at least as important as *The Chaos Scenario* itself — which is why the 12th chapter of this book is available only at TheChaosScenario.net (first in blog form and later to be fleshed out) documenting our successes, our misses, our humiliating pratfalls. I can say categorically that you have not finished this book until you have taken in the website. It is astonishing.

Naturally, I hope this book is read by every man, woman and child on the face of the earth, and that the Stielstra imprint prospers far beyond this project. But however this all pans out, the exercise has been the most exciting undertaking I've ever participated in — and I've bowled in the White House, played high-stakes poker, bungee jumped, eaten dog in Seoul and attended *Cats* unmedicated.

Beyond my humble labors, *The Chaos Scenario* resulted from enormous contributions from a vast array of friends, colleagues and major offspring. The entire *Advertising Age* editorial staff has been unstintingly cooperative, most especially Abbey Klaassen, Jack Neff, Kevin Brown, Brian Steinberg, Michael Learmonth, Nat Ives, Matt Creamer, Rupal Parekh and Jesper Goransson. My editors Ken Wheaton and Jonah Bloom offered more freedom and wise counsel than I could ever have wished. Crain Communications executives David Klein, Allison Arden and Rance Crain have supported me every step of the way — even when I bit the hands that feed them. Scott Donaton was present at creation on this project, and invested a whole lot of editorial capital to make it happen.

At WNYC, the *On the Media* staff was no less important, as I cagily steered my story selections to parallel the book — making a whole team of producers my *de facto* research team. As they are the smartest group of people I've ever worked with, they were on to the gambit from the get-go, yet they played along cheerfully and brilliantly. Thanks to Megan Ryan, Nazanin Rafsanjani, Mike Vuolo, Mark Phillips, Jamison York, Tony Field, P.J. Vogt, Michael Bernstein, Nadia Zonis and, on the engineering side, Jennifer Munson and Dylan Keefe. The management team of Katya Rogers, John Keefe and Dean Cappello offered the most generous encouragement at every stage. As for Brooke Gladstone, my co-host, editor, mother confessor, dear friend and arch-nemesis, I wrote this book mainly to impress her.

At *Wired*, I offer my gratitude, appreciation and endless admiration to senior editor Mark Robinson, who shepherded me through my original YouTube reporting, and to Chris Anderson, a visionary editor and writer who I feel privileged to have a) worked with, and b) strong-armed a blurb from.

A couple of hundred interviews are the armature upon which the whole text hangs, but some of my sources were heroically helpful: Jessica Greenwood, Ayelet Noff, J.D. Lasica, Jeff Jarvis, Jeff Howe, Rishad Tobaccowala, Seth Godin, Clay Shirky, Ted Leonsis and most of all Neil Perry, who opened up his shop to me — in good times and bad — at incalculable risk to himself. My friend and mentor Steve Rosenbaum was my sounding board for the entire duration of this project. My friend (and former editor) Linton Weeks read my raw manuscript and, as always, offered unerring guidance. My daughter Katie Whitehill went over the final version with a fine-tooth comb, yielding (I'm embarrassed to say) more than 200 corrections and many thoughtful suggestions for improvement. She called me awful names here and there, but, in the spirit of my subject matter, I listened.

My thanks to my agent Dan Strone for his indefatigable efforts, to Bart Wilson, for his labors on ComcastMustDie.com, to Jeffrey Buntin Sr. and Jr., for giving me access to their digital brain trust in the person of Greg Stielstra and, of course, to Greg himself. My publisher and business partner in this venture is a man of boundless knowledge and energy, not to mention strategic thinking and an overflowing cornucopia of ideas. If this book is successful, I assure you, the credit belongs as much to him as to me.

Finally, a word about the title. Why name the book *The Chaos Scenario* when 75% of it is about not the problem but the solution? Excellent question. When you're writing about control of our day-to-day life transferring from institutions to individuals amid the transformation to a digital world, all sorts of ideas suggest themselves. For instance: *The Mouse That Roared*. Yes, that was the name of Leonard Wibberley's 1955 comic novel about a tiny, backwards European duchy that declared war on the United States ... and won. But of course it also describes the ascendancy of the single computer user over the erstwhile Powers That Be. Alas, a duly diligent Google search revealed that at least one marketing consultancy had alighted on the name notion, so that one was out the window.

For a slightly different reason, I discarded another literary reference.

Lilliputt seemed just so perfect, because of how those tiny little people were able to subdue the giant Gulliver. But once you get beyond that, the Jonathan Swift satire was about something else altogether: namely two dysfunctional societies warring over the most idiotic of differences. Try as I might, I couldn't spin his theme to conform to my premise.

Death to Pluto amused me because it referred to the demotion in the celestial hierarchy of an entire (former) planet. If the roster of the solar system is subject to revision, seems to me, no reason to think that, say, network TV is somehow immutable and eternal. For similar reasons, I gave a fair amount of thought to *The King is Dead. Long Live the King* and *The End of Everything (And the Beginning of Everything Else.)*

For the first year, the working title was *Listenomics*, but, as I mentioned in the introduction, *Wikinomics* and *Freakonomics* beat me to press (and since then *Womenomics*, too). Still, I had plenty of other options: *True Tales of the Mob, The Guillotine and the Scalpel, King You* and, (my particular favorite) *Today's Bad But Let's Try to Connect After the Apocalypse.*

Penultimately I seized upon *Name This Book*. Now that was a cool one. If the whole point of this exercise is to explain and dramatize the ways in which the power pyramid has been turned upside, maybe it would've been silly of me — and maybe even arrogant and hypocritical — to presume to define for you what it all means. If my premise in spending four-plus years writing this thing is that it is increasingly fruitless for The Man to dictate to the peeps, the least I could do was practice what I preach. In fact, apart from the title, the whole production, marketing and distribution of this book cleaves to the principles discussed herein. In further fact, *you're not even reading the finished product*, because the final chapter — available only online — will chronicle how all that stuff was done, and how it turned out. So, it would have been just so perfect to say, "Call this baby anything you like. The matter is literally in your hands."

But my wife said that was stupid, and I have to live in this house. Hence, a final decision based on what they call "brand equity." *The*

Chaos Scenario was the title that launched this whole endeavor. It has become a sort of term of art in the marketing world. They do say to stick with your first SAT answer. And, hell, it looks fantastic on a book cover.

— Bob Garfield
Potomac, Md
May 2009

ABOUT THE AUTHOR

Bob Garfield isn't exactly a media whore, but he's extremely promiscuous.

For 24 years has worked for *Advertising Age*, where his ad-criticism column has made him an institution, like the Red Cross. Or San Quentin. He is also co-host of NPR's *On the Media*, a cleverly-titled radio show that reports on the media. In both gigs he has won many journalism awards, several of them doozies.

Bob is a founding contributor to the Watchdog Blog of the Nieman Foundation for Journalism at Harvard University. He's been a contributing editor for the *Washington Post Magazine, Civilization* and the op-ed page of *USA Today*. He has also written for *The New York Times*, *Playboy, Sports Illustrated* and *Wired* and been employed variously by ABC, CBS, CNBC and the defunct FNN as an on-air analyst. As a lecturer and panelist, he has appeared in 30 countries on four continents, including such venues as the Kennedy Center, the U.S. Capitol, the Rainbow Room, the Smithsonian, Circus Circus casino, the Grand Ole Opry, the U.N. and, memorably, the Westward Ho! motel in Grand Forks, N.D.

His first book, *Waking Up Screaming from the American Dream*, was published by Scribner in 1997, favorably reviewed and quickly forgotten. His 2003 manifesto on advertising, *And Now a Few Words From Me*, is published in six languages (although, admittedly, one is Bulgarian). Garfield co-wrote "Tag, You're It," a snappy country song performed by Willie Nelson, and wrote an episode of the short-lived NBC sitcom *Sweet Surrender*. It sucked.

INDEX